THE BIG FIX

THE BIG FIX

THE HUNT FOR THE MATCH-FIXERS BRINGING DOWN SOCCER

BRETT FORREST

wm
WILLIAM MORROW
An Imprint of HarperCollins*Publishers*

HarperCollins books may be purchased for educational, business, or sales promotional use. For information please e-mail the Special Markets Department at SPsales@harpercollins.com.

A hardcover edition of this book was published in 2014 by William Morrow, an imprint of HarperCollins Publishers.

FIRST WILLIAM MORROW PAPERBACK EDITION PUBLISHED 2015.

Designed by Jamie Lynn Kerner
Photograph on title spread © by Helder Almeida/Shutterstock, Inc.
Soccer player on chapter openers © by Christos Georghiou/Shutterstock, Inc.

Library of Congress Cataloging-in-Publication Data has been applied for.

ISBN 978-0-06-230808-5

15 16 17 18 19   OV/RRD   10 9 8 7 6 5 4 3 2

For Cindy and Eric, who got me back in the game

CHAPTER 1

KHALID BIN MOHAMMED STADIUM
SHARJAH, UNITED ARAB EMIRATES, MARCH 2011

The FIFA operatives arrived at the stadium past noon, prepared to disrupt the crime that was tearing soccer apart. Sharjah was a short drive up the road from Dubai, but it felt like a world away, in the unglamorous dust, a face of the United Arab Emirates that most Westerners never saw. Unlike Dubai, Sharjah didn't look like a place where someone could get rich overnight. This made it a fitting location for the criminals who had infiltrated the game of soccer. Their specialty was illusion, and events in Sharjah were about to give them another lucrative ninety-minute return.

The match set for that day—March 26, 2011—was an exhibition, or friendly, between the national teams of Kuwait and Jordan. It was the sort of game that took place hundreds of times each year around the world, with limited notice and consequence. National team coaches often looked upon these contests as little more than vigorous practice. Interested criminal groups from Southeast Asia to Eastern Europe, on the other hand, considered them the cornerstones of an extensive commercial enterprise.

Now Kuwait versus Jordan was forming into the front line in a gathering war. On one side were organized criminal syndicates, which were making hundreds of millions of dollars—if not billions, the total amount a drop in the opaque, trillion-dollar pool of soccer betting—by manipulating game results. On the other side were soccer administrators, who were beginning to accept that match-fixing was the stunning sports scandal of our time, a fundamental threat to the most popular game in the world.

FIFA (Fédération Internationale de Football Association), soccer's international governing body, had received information that a collection of known criminals had arranged to manipulate the outcome of the Sharjah match. This was no great surprise, as inflated final scores, questionable penalty calls, and strange betting patterns had been occurring with great frequency in recent seasons. What was novel about this Sharjah match was the fact that FIFA, directed by a newly hired security chief, was operating a clandestine investigation. The time had come to take action.

When the two FIFA investigators entered Khalid Bin Mohammed Stadium, shortly before game time, there was no one else there. It had been a challenge to gather reliable information about the match, even for FIFA, which had sanctioned the competition. The date, the kickoff time, the location—all were in doubt. The websites for the Kuwaiti and Jordanian teams provided conflicting information. Gambling websites did as well. Some sources even listed the match as having been canceled. That's what it looked like to the two FIFA men as they passed through the wide-open stadium gates. Nobody was selling tickets. The stands were empty. The FIFA investigators took seats in the grandstand, and they noticed that there were no TV cameras or production vehicles present. The match had not been advertised in the local press. In an age of constant coverage and information, it felt like this game was going to take place only in the imagination.

Eventually, the players filtered into the stadium, as did a few fans. Several men milled about on the edge of the playing field, and the FIFA operatives took particular notice of them. They recognized the men, one from an Emirati promotional company, another from a similar firm in Egypt. They had helped arrange the match, though they were ultimately incidental to FIFA's investigation. Instead, FIFA was interested in the architects of this fix, a notorious group from Singapore, which operated unchecked in dozens of countries across the world. The FIFA investigators noticed two known Singaporean fixers enter the stadium and take seats in the VIP stands. The match was set to begin.

The Sharjah fix had originated in the mind of the most prolific match-fixer in the world, a man of mysterious movements who had manipulated hundreds of matches in more than sixty countries, making untold sums for the syndicate in Asia. But the syndicate had betrayed him. Police had discovered details of the Sharjah fix sketched on a piece of paper lying on his hotel room bed in a Finnish town along the Arctic Circle.

This information had led FIFA's investigators to Sharjah. They planned, at halftime, to barge in unannounced to the locker rooms of the players and officials, threatening suspension—or prosecution—unless the second half was contested honestly. But although the two FIFA men had attempted to contact officials from the UAE Football Association, their calls and emails had gone unreturned. For now they were relegated to the stands, without the proper credential to visit other areas of the stadium. They speculated that the UAE soccer officials may have knowledge of the upcoming fix. Plenty of national soccer federations around the world had gone into the lucrative business of fixing with the Singapore syndicate itself.

The purpose of match-fixing was betting fraud. Fixers compromised players, directing them to allow the other team to score. Fixers bribed referees to hand out red cards and penalty kicks, thus influencing the outcomes of matches. The syndicate placed bets based on the timing of these planned events. The fixers defrauded both the bookmaker, who was always one step behind them, and the fan, who believed that what he was watching was real. There was also the player, often coerced into participating. When play began between Kuwait and Jordan, activity on the international betting market revealed that the fix was in.

At some point in the 1990s, Joseph "Sepp" Blatter, the president of FIFA, began characterizing the numerous soccer players and administrators around the world, collectively, in his public comments, as the "football family." FIFA, headquartered in Zurich, is the organization that is responsible for staging the World Cup every four years. Among the patchwork of federations and confederations that control and administer soccer, FIFA carries the most weight. It is the organization that most soccer people petition in order to solve a dispute, or to dispense a handout. But FIFA is hardly the guardian of goodwill that Blatter's benevolent terminology suggests. FIFA is registered in Switzerland as a charity, yet it does not operate as a traditional nonprofit organization, with its income of $1 billion per year and its multifarious corporate sponsorship deals and TV contracts. Nor does FIFA behave like a modern business, with standard corporate checks and balances. Instead, it resides somewhere in the imprecise middle, and for some of its top executives this is just fine, ambiguity being the facilitator of exploitation.

In the last decade, the imprecise way that global soccer is administered has exposed it to crisis. Match-fixing has overtaken the sport. It is not the fault of FIFA that international organized crime has targeted soccer. But considering the criminal nature of match-fixing, Blatter's words have assumed a new meaning. This is soccer's modern "family":

Operation Last Bet rocked the Italian Football Federation, as fifteen clubs and twenty-four players, coaches, refs, and officials were implicated in match-fixing. Turkish police arrested nearly one hundred players, while the Turkish Football Federation excluded its club, Fenerbahce, from the UEFA Champions League, questioning how the team managed to win eighteen of its last nineteen games to take the domestic title. The Zimbabwe Football Association banned eighty players from its national team selection based on suspicion of match-fixing. Lu Jun, the first Chinese official to referee a World Cup game, was jailed for five and a half years for taking bribes totaling $128,000, enhancing the meaning of his nickname, the "golden whistle."

In South Korea, prosecutors charged fifty-seven people with match-fixing; two players subsequently committed suicide, rather than face the shame. Two Brazilian refs were handed prison terms and the Brazilian Football Confederation was fined $8 million for their roles in a series of fixes. As soon as eight Estonians received a one-year ban, a court charged another dozen with fixing. German police recorded Croatian criminals discussing by phone their plans to fix games in Canada. The disgraced head of the Chinese Football Association is currently serving time in a penal colony for match-fixing. Hungarian police arrested more than fifty people for fixing, though before they could apprehend the director of a club under scrutiny, he jumped to his death. The Czechs are prosecuting two referees for fixing. The Cambodian national team manipulated its failed two-game series with Laos, a qualifier for the 2014 World Cup.

Macedonia is so corrupt that few bookies will take a bet on a game in its domestic league. The executives at a Bulgarian club, Lokomotiv Plovdiv, ordered their players and coaches to take a lie detector test after a loss. Georgian players, team owners, and bookies are behind bars for match-fixing. In Malaysia, a few dozen players are currently in custody. Kenyan, Lebanese, and Tanzanian refs have worked fixes. Niger bred the most corrupt referee of all. Polish authorities have prosecuted a dozen players for fixing. The Russian government has established a committee to eradicate match-fixing from its leagues. The prime minister of Belize ordered a match-fixing probe into the head of the country's soccer association.

Chinese and Italian organized crime have targeted the Belgian league for years. Bosnia's league is targeted by Bosnia's criminals. Switzerland banned nine players for fixing. Italian prosecutors pinned fixing charges on Gennaro "Rino" Gattuso, a popular, World Cup–winning midfielder and former star for AC Milan. Gattuso said he was prepared "to go into the town square and kill myself in front of everyone if I should be found guilty of such a crime." Two scandals rocked English football this past fall, one involving Singaporean fixers, the other ensnaring a former Premier League player. Germany is prosecuting the most famous fixing case of all, in Bochum, which has revealed that a network of criminal match-fixers has been manipulating soccer in every corner of the world for the better part of the last decade.

Could the state of affairs be this bad? Unequivocally, yes. Today, there are active police investigations into match-fixing in more than sixty countries, which is about one-third of the world. Half of the national and regional associations affiliated with FIFA have reported incidents of fixing. One can only imagine the amount of fixes that have transpired with only the perpetrators in the know. The fixing of international soccer matches has become as epidemic as drug trafficking, prostitution, and the trade in illegal weapons. This is happening in a sport where the players walk from the locker room to the field hand in hand with small children, as though soccer were a refuge of innocence and moral purity. Overwhelming evidence presents a contradictory argument: the most popular game in the world is the most corrupt game in the world.

Gambling is to blame. The market in sports betting has ballooned in the last decade, its illegal portion rivaling long-established criminal enterprises. Interpol claims that $1 trillion is bet on soccer games per year. Asian bookies suggest a much higher figure. The soccer industry itself—the TV contracts and sponsorship deals that comprise the business of the game—is estimated at an annual value of just $25 billion.

Unpoliced, driven by easy profits, match-fixing has grown out of control. Superior clubs lie down for inferior clubs that are trying to avoid relegation to a lower division of competition. Coaches, players, refs, and government officials collude for the fix. International qualifiers result in outrageous scores: 11–1, 7–0. The opportunity for easy profits drove early, creative attempts to manipulate results. On November 3, 1997, in an English Premier League match against Crystal Palace, West Ham scored to tie the game at two in the sixty-fifth minute. Abruptly, the stadium lights went out. The same thing happened when Wimbledon played Arsenal a month later. A Chinese-Malaysian gang had paid the technicians at the stadiums to cut the power when the game had reached the desired score. Enveloping greed has caused the players themselves to take severe measures to enforce the fix. In a 2010 Italian match, a goalkeeper allegedly drugged his own teammates at halftime so that opponents could easily outrun them.

The players are inconsequential. They are tools of the syndicate bosses who operate in the shadows. For the established criminals, international soccer has been a free zone of activity, a territory of endless opportunities for manipulation. Each of the nearly two hundred countries that FIFA recognizes has a professional league and a national team, which is classified into several age groups. The total worldwide number of national and professional soccer teams exceeds ten thousand. Multiply this figure by the number of players per team, then add referees, club officials, and federation administrators, and the entry points for a match-fixer are voluminous and as ever changing as a club roster from season to season. There is no centralized control, no disciplinary commissioner. International soccer is a loosely administered network of different languages and customs and laws and economies and currencies that connects the world, yet barely hangs together. This variance gives the game its special charm. It also allows dark motivations to flourish. Criminal fixing syndicates have infiltrated the game of soccer so fundamentally, manipulating the betting market to their advantage, that they have called into question the outcome of every match in the world.

The opening moments of the Kuwait-Jordan match elapsed at a brisk pace. There were several heavy tackles. A man sitting behind the FIFA investigators laughed, remarking that people from Kuwait and Jordan disliked each other. The referee made a questionable penalty call in the game's twenty-third minute, when the ball ricocheted off the hand of an unwitting Jordanian player. Kuwait converted. The FIFA operatives looked at the Singaporean fixers in the crowd, but their body language revealed nothing. It didn't have to. The evidence was in the numbers.

There are several ways to fix a match. One of the most popular is to wager on the total number of goals scored. If a bookie lists the over-under at 2.5, and a fixer bets on the over, he will direct his compromised players or referee to make certain that three goals or more are scored in the game. If he bets on the under, then he will order two or fewer goals.

The fixing syndicate operated on the in-game gambling market, which allowed for betting as a match progressed. At the opening of the Sharjah match, 188Bet, one of the largest bookmakers in the world, began taking a preponderance of bets that supported three goals or more. The 188Bet odds for three or more goals started at 2.00, or a 50 percent probability. At the match's eighteenth minute, the match still scoreless, the odds for three or more goals decreased to 1.88, or a 53 percent chance. These figures revealed a telling detail. At the beginning of the match, with ninety minutes in which to score three goals, 188Bet calculated the chance of three or more goals at 50 percent. Paradoxically, after eighteen minutes had elapsed, the chance of three or more goals was now greater, even though there was less time—only 80 percent of the match remained—in which to score them.

The bookies at 188Bet had not determined that an outcome of three or more goals was now more likely. What they had done was moved the odds in reaction to the overwhelming wagers they were receiving on three or more goals. It is the bookmaker's goal to level his book, taking equivalent action on either side of a proposition in order to reduce his exposure and profit from his margin. And a bookie knows that his exposure is highest when he is taking a large amount of action on an illogical proposition. He knows that the match is fixed, as the bookies at 188Bet surely understood as they calculated in-game odds for the Sharjah match.

As the game neared halftime, only one goal had been scored. In the thirty-eighth minute, the referee called another penalty. This one appeared legitimate, as a Kuwaiti defender tackled a Jordanian forward in the box. The goalkeeper for Kuwait saved the ensuing kick. However, the linesman flagged him for early movement. Jordan scored on the retake. At the half, the match was tied at one. With forty-five minutes to play, all the syndicate needed to win its bets was one more goal. Easy. But then something happened.

In the grandstand, the FIFA investigators contemplated attempting to bluff their way into the locker rooms to confront the referee and players. As they did so, they watched the man from the Emirati promotional company climb the stairs of the VIP stands. He spoke with the Singaporean fixers. As FIFA later discovered, the match referee had received a tip that the contest was under observation. The players returned to the field for the second half, and the Singaporeans left the stadium. Midway through the second half, the score was still 1–1.

Suddenly, in the seventy-first minute, the betting action reversed. There was no longer heavy action at 188Bet for more than three goals, although there were nineteen minutes remaining in which to score for the third time. Warned that FIFA investigators were present in the stadium, the syndicate had pulled the fix, pulled its bets. The match finished in a 1–1 draw.

From the chatter that FIFA's security team subsequently collected in Singapore, syndicate members were confused, wondering who had leaked word of the Sharjah operation. The syndicate had lost roughly $500,000 on the Sharjah match, according to FIFA intelligence. Considering the size of the soccer betting market, this wasn't a considerable number. However, the event in Sharjah was significant. No one had ever fought the syndicate before. Sure, there had been traditional prosecutions, investigations conducted after the crime had been committed and the profits earned. But never before had FIFA conducted a counter-fixing operation in real time. Asian fixers and their European partners had operated freely for a decade. Everything was about to change.

CHAPTER 2

SINGAPORE, 1983

The best soccer players are poor. Wilson Raj Perumal came to this understanding decades ago, sitting in the grandstand of Singapore's Jalan Besar Stadium. It was the mid-1980s, and the old grounds, near the city-state's Little India neighborhood, were hosting a domestic league match. Perumal had no rooting interest. For him, it would always be about the odds, the line, the payout.

During the World War II occupation of Singapore, Japanese authorities set up camp at Jalan Besar, where they culled the Chinese from the population, marking them for summary execution. Now, to Perumal, it was the Chinese who stood out most of all. Sipping tea, betting on matches (contrary to local statutes), they orchestrated the action that worked Perumal's mind toward opportunity.

Perumal had plenty of friends who played organized soccer. He understood the game. What he didn't understand was how these old men had been taking his money for the past half year. Perumal had started betting for fun. It was something to do with his friends from school. He didn't comprehend what was known as *hang cheng* betting, which was determined not by which team won the game, but by an agreed-upon value of one's bet. The line dictated the score by which one team had to win, and the odds corresponded to the likelihood of that event coming to pass, establishing the amount of a winning bet. Once he learned this, Perumal easily recognized the pattern of his losses. Each time Perumal placed a bet, the Chinese men changed the odds and the line to suit themselves. They had been manipulating the market depending on which team Perumal chose, which bet he wanted to place. They had been rigging the action all along. Determined on winning back the money he had lost, and more besides, Perumal got an idea.

Theft was the first charge the cops hung on Perumal, for stealing a pair of soccer cleats. This was in 1983. Eighteen years old, Perumal lived with his family in Choa Chu Kang, a farming area in Singapore's northwest. His parents traced their roots to Chennai, the capital city of India's heavily populated Tamil Nadu state, a wellspring of cheap labor to Asia and the Middle East. They were part of a long line of convicts and unskilled workers who had made their way from India to Singapore during the century leading up to World War II, when the two territories existed under British colonial governance. Perumal's father, a simple laborer who painted street curbs and laid cable, was a black belt in judo. Perumal never took to such discipline.

Instead, the lasting impression that his father gave him was how difficult it was to feed five children on honest industry. "Some days we had to make it on one meal," Perumal says. He was the middle child, the one who gets lost in the shuffle, the one who finds other ways to survive. He attended Teck Whye, the local school where he ran the 800 meters and paid passing attention to his studies, more interested in the dubious extracurricular activities that awaited him after school.

Perumal and the Singaporean state were born the same year, 1965, though their characters instantly diverged. Upon exiting the British realm, Singapore's leaders placed the country on a path of economic vigor. Shipping, manufacturing, and industrialization transformed Singapore into one of the four Asian Tigers, a center of international business and finance. Underlying this growth was Singapore's commitment to discipline, its blunt approach to crime. Unlike many of its neighbors, stricken with the chaos of liberalism or the stagnation of autocracy, the Singaporean city-state struck a balance: tough on crime, friendly to business. Singapore became a place where the sinner was punished disproportionately to his sin, so that the innocent could prosper likewise beyond proportion.

Wilson Perumal belonged to the third-largest ethnic group in Singapore. There was no such thing as a Singaporean. There was Chinese, Malay, Sinhalese, Filipino, Thai, each with a different language, each adopting English as default dialect, each keeping secrets in their particular tongues. Singapore was a place of secondary identities, a place of no insiders. Perumal skimmed from one social set to another, between ethnic groups, learning to conceal his motivation in order to persuade and gain advantage. He could have made an effective salesman if he hadn't gravitated to kids who likewise couldn't conceive of their futures, just what they could get their hands on right now. Perumal tried his hand at petty crime. With a few friends, he stole a VCR from the Teck Whye School. They sold it, pulling in five hundred Singapore dollars. They took a cab downtown, saw a few movies, blew the money on popcorn and beer, incautious about what they had done.

Later, a member of his crew stole a pair of soccer cleats, and this led to the group's undoing. Confronted by the authorities, the friend told the whole story, about the shoes, the VCR, and other thefts, implicating Perumal in so doing. The next day, a headline in the local paper read: "Asian School Athlete Charged with House Breaking." It was the sort of teenage troublemaking that often scares an adolescent onto the right path in life. In Perumal, the episode simply provided his first publicity. There would be much more.

Perumal was now acquainted with criminality, yet this was hardly the most serious offense in Singapore. The country had become a disciplined, transparent economic model for the world, yet illegal betting remained the most tolerated crime there was, a clandestine element of the culture. There was little the strict government could do about it. Everyone gambled. Just as Perumal did, at Jalan Besar Stadium.

When Perumal realized that the Chinese men had taken advantage of him, he turned his attention to the players who sprinted and struggled in the clinging Singapore humidity, less than one hundred miles north of the equator. Perumal knew what that was like, to work hard for little reward, growing up with nothing in your pockets, with few prospects to fill them, your restless energy leading in self-destructive directions. Perumal understood the point of developing a singular focus on something that might carry you out of poverty. Along the way toward on-field glory, he thought, what was wrong with making a little something on the side?

He related this reasoning to several of his friends who played soccer. Everyone saw eye to eye, common understanding being the essential element of manipulation. He purchased two sets of soccer jerseys. One red, one white. He rented a local stadium, paying a hundred dollars to monopolize the field for two hours. He listed the match in the local papers. He bought a pair of shorts, a polo shirt, and socks and shoes—all black—draping them on a friend. "You're the referee," Perumal told him.

When the Chinese bettors from Jalan Besar Stadium, always looking for action, read about the match in the newspaper, they showed up at the appropriate time and location. When the red team went up 2–0 at halftime, the old Chinese men were all too happy to bet on red to win the match, handing their markers to Perumal and smiling to themselves at the kid who didn't understand *hang cheng*. When the white team had scored its third goal of the second half, the old Chinese men weren't laughing anymore. They knew that the kid who was learning the ropes had just roped them into a scam.

Perumal had found his calling: easy money. His first fixed match was so successful that he carried it out in stadiums throughout Singapore. The losing bettors didn't complain, even though they sensed something tricky about these wagers and these games. They couldn't go to the cops. They couldn't grouse and lose face. All they could do was pay Wilson Perumal what they owed him.

Perumal pursued this scheme into his twenties, and he developed a taste for things that he could never have before. It was the first time he had any money. Running through pool halls and chasing girls with his friends until the sun came up, he bet his earnings on matches in Europe's biggest soccer leagues, the matches that were just starting to be televised in Singapore. As he watched the games, in that charged, early-morning condition of fatigue, youth, and stimulation, Perumal conceived of something bigger.

CHAPTER 3

With a mustache that runs long and tall and out of date, Chris Eaton calls to mind a frontier sheriff, the one man willing to establish justice on the range, where the sun catches his tin star, confirming the higher calling of order. "I quite would have liked that," Eaton says. "There are a lot of people that need shooting on the edge of the corral."

Eaton comes at you with the inevitable momentum of an arrest. However, his Australian informality requires you to remind yourself that he is an upholder of the law. Sixty-two years old, Eaton has the energy of a thirty-year-old. He has fathered six children, the youngest now just two years of age, confirming that he's not slowing down. "Life is for living," Eaton is fond of saying. "Not for rule-making." His firm moral foundation, however, is a touchstone shallowly concealed, a lager in hand all that's necessary to lead him sometimes to soliloquy.

The speech he gives these days invariably instructs the ill-informed, the morally lax, and the financially curious about the inner evils and workings of match-fixing. In European conference halls and Asian banquet rooms and the New York bar or two, Eaton arrives as featured speaker, the face and voice of the fight against "the manipulation of sporting events for the purpose of illegal betting." He is an official carved perfectly to combat fixing. Eaton is dogged, antipolitical, rule-bound, perceptive of people, and not afraid of an audience, which he doesn't coddle. "Chris talks to powerful people like they've never been spoken to," says one of his lieutenants.

When Eaton leaves these powerful people—elected officials, police superintendents, administrators in the sporting world—they often shake their heads in derision. *Match-fixing could never happen to us.* Invariably, months or maybe a year or two later, when enough time has passed for Eaton to fade from their thoughts, suddenly he returns. What he predicted has come to pass. And he is the only one to call for help, because no one else knows what to do. This has happened so often as to defy coincidence. The billions of dollars available in the manipulation of soccer matches are too tempting for organized crime to ignore, and match-fixing creeps into every local market. Eaton spreads his gospel and combats his criminal opponents, his monthly itinerary a checkerboard of takeoffs and landings from one continent to another. A lifelong policeman, he has become soccer's redeemer, the one man with the will and the strategy to scuttle match-fixing and restore the integrity of the game.

Eaton never wanted to be a cop. In his view, the police force was a destination of lowly ambition. In 1960s Australia, it was.

Policing employed muscle, rather than cunning. It reflected not only the predominate domestic view, that the law cast no shades of gray, but also the country's sporting culture. Australian rules football was the sport of choice, a game that developed a man's ability to wear down his opponent in barely legislated brutality. Soccer, the thinking man's game, a sport of deft artistry, was the province of European émigrés, awkward souls stranded Down Under who gathered now and again on the patchy turf of neglected fields, communicating in this foreign language of strategy.

It was Eaton's older brother, Ian, the firstborn of the family, who wanted to wear blue. At eighteen years old, he was the right size, six foot two and 200 pounds, big and rangy enough to succeed with aggression in the Victoria Police, but he failed the police exam.

Life's path navigated away from Ian's control, while Chris was certain that he would draw his own. The family spent its Christmas vacations at Mornington, outside Melbourne, bunking together in a mobile home, where Chris would break out pencil and paper. He had inherited a talent for sketching from his father, an architect, who encouraged him toward the profession. But Chris was interested in the human form. While he sketched the outlines of a face or a torso, he felt a person take shape in his understanding—how a well-placed stroke could manipulate them to one position or another. He thought that he would attend art college in Melbourne.

Ian's path carried him to the army, though it always meandered back to Mornington every summery December. Ian would pack into a car with two of his friends, headed for nearby Cape Schanck, where the nineteenth-century lighthouse brought in the tourists, while the girls in bikinis attracted their own local attention. One clear afternoon, the boys wandered along the cliffs that overlooked the beach, the waves elapsing along the rocks, and the sandy pathway crumbled underfoot. Ian fell the full seventy feet to the rocks, causing the brain hemorrhage that killed him.

Sixteen-year-old Chris watched his mother sink into depression. Regret consumed his father, a career man who had known his oldest son only passingly. Chris's younger brother, Anthony, was neglected. Chris put away his pencils and drawings, and he abandoned school in favor of the police academy. This would be his way of memorializing Ian. He didn't see himself as a cop, but by sacrificing himself so that his family might emotionally recover, he displayed the character of the ideal policeman— shielding the victims, even if he didn't realize, in his youth, that he was a victim, too.

At Melbourne's St. Kilda precinct, Eaton looked the part physically, like his brother big enough to handle himself. But the difference in temperament between his colleagues and himself was so striking that Eaton was certain he was in the wrong profession. By the 1970s, St. Kilda's nineteenth-century seaside mansions had been sectioned into apartments for low-income families, and the neighborhood became a dim environment of drugs, violence, and prostitution. Crime was such a part of life in St. Kilda that police could apply no lasting solution to it. They could only identify "natural criminals," night-sticking them into temporary submission. "We were really the thin blue line in those days," Eaton says. "I learned quickly that policing was there to repress the troublesome in society from those who didn't want to be troubled by them."

This was no element for the righteous or the philosophical, or even the merciful. One afternoon, police detained an offender, who arrived at the St. Kilda precinct. The man had groped a girl on the beach, but by law Eaton had to let him free. There was not enough evidence. Eaton later learned that the man had gone on to rape and murder. And so while the roughhouse nature of St. Kilda policing offended Eaton's cerebral disposition, experience broadened his view. "By taking no action, you exude weakness," he says. "Criminals only respect authority. And authority doesn't come from the uniform. It comes from a style."

In a place beholden to gangs, the police were St. Kilda's biggest gang of all. As Eaton looked through the bars of the precinct's back window and out onto Port Phillip Bay, he realized that he hadn't signed up to be part of a posse. Each night, he felt for a solution, as he transited from the charged environment of the St. Kilda streets to his wife, Debbie, back home.

Debbie was the slim, brown-haired girl next door, laughing and animated. She was also sixteen, and the two married in a shotgun wedding in 1972 when Chris was nineteen and a rookie in the Victoria police force. They named their son Ian. A daughter, Sarah, came along in 1976.

As if compensating for the education that he had relinquished, Eaton became a reader of great hunger and interest. In the pages of the books that he read in his young family's two-bedroom apartment, he encountered mention of an organization that might serve as a model for his own. It was the FBI's cerebral approach to crime prevention that agreed with the ideas Eaton was rapidly developing. He admired the work of J. Edgar Hoover, if not the man himself, and Hoover's vigorous application of the law to the influential, whereas police had before applied it only to the impoverished. In Australia, Eaton saw a mirror image. "The people who were committing the big crime in Melbourne, the people with money, the people who were committing enormous frauds on society, police didn't even pay a note's attention to them," he says. Eaton understood that the crime that was visible on the streets of St. Kilda was the result of greater forces, grand manipulators hidden from view. He realized that it wasn't enough to cultivate authority, but to apply it to effect.

Eaton wrote about his progressive ideals in the police journals. This gained him notice and promotion into the Australian federal police, working in Canberra, the Australian capital. Not yet thirty years old, Eaton had achieved all of the things that his brother Ian had hoped he would in a lifetime.

He was enjoying a cool respite in 1981 as he steered his Ford Fairmont north along the M31 highway, on his way home from Melbourne, where he had just served as best man at the wedding of his brother Anthony. The kids were asleep in the backseat. Debbie was slouched against a pillow in the passenger seat, her eyes closed. The fog in the air wisped in spirals around the rushing frame of the Fairmont coupe. It wasn't a long trip from Melbourne to Canberra—five hours if you drove like you meant it—and Eaton was taking it slow. There was no rush. He steered along the highway's winding curve, enjoying the way that felt, to be in control. Headlights roused him from his thoughts.

CHAPTER 4

KUALA LUMPUR, 1990

Rajendran Kurusamy would stride into the raucous stadiums of the Malaysia Cup like he was the tournament commissioner. In many ways, he was—controlling which players saw the field, determining winners and losers, paying referees and coaches from an ever-renewing slush fund. The Malaysia Cup was a competition between teams representing Malaysian states, along with the national teams of Singapore and Brunei. It was the early 1990s. Talking on his clunky, early-model mobile phone, Kurusamy would attend a game long enough for the players on the field to notice that he was there, remembering the money they had taken from him, understanding that the fix was on. Kurusamy would leave the match as forty-five thousand fans celebrated a goal, unaware of the man who had set it up.

Those who knew him called him Pal. Those who made money with him called him the Boss. Those who owed him money often didn't have the opportunity to call him anything at all, Kurusamy's muscle engaging in one-sided conversations. Kurusamy was the king fixer in the golden age of the pre-Internet racket.

As Kurusamy walked out of Stadium Merdeka, with its view of the Kuala Lumpur skyline, Wilson Perumal was just walking in. The Petronas Towers were elevating into the sky, soon to be the world's tallest buildings. Perumal was also rising in the estimation of those around him. His Chinese contacts from the small-time Singapore action respected him for the lumps he had given them. They pulled him along to the livelier action of the Malaysia Cup.

The betting was heavier than anything Perumal had ever seen. Men who displayed no outward signs of wealth would bet $100,000 on a game, and more. It was a frenzy, the action conducted through a web of runners and agents who transferred bets to unseen bookies. Chinese, Malaysians, Indonesians, Thai, Vietnamese. You called and placed bets over the phone. You had to build up a reputation before a bookie would take your bet, but it all happened quickly, as long as you paid your losses. No one knew who sat at the top, who pulled the strings, just that the bets escalated higher and higher, and if you delayed in paying a debt, it wouldn't be long before someone paid you a visit. This was the action that Perumal had been looking for, and he fell to it naturally, any thought of a conventional life left behind. "If I go to work for thirty days, I earn fifteen hundred dollars," he said to himself. "But here, I am gambling fifteen hundred per game. It doesn't tally." His wins got bigger, but his losses did, too. The point was that his money was in motion, which was a trait of a high roller, the only person Perumal wanted to become. He looked around, and he realized as the games played out on the field that there were no fans, just bettors. The match was a casino. The players were the dice or the cards, which could be loaded or marked by the manipulators who gravitate to apparent games of chance.

The games of the Malaysia Cup were not games of chance, or so the chatter led Perumal to believe. In the stands or on the phone or on the street, he would hear of the fix. Few people knew for sure. But everybody could tell. Perumal watched the ripple cascade through the ranks of the bettors, and he recognized the real game and who possessed the power in it. He learned to take advantage of the hints he heard, throwing his money in the direction of the fix. As he collected his winnings, he heard the name Pal. If you could get close to Pal, people said, you would know which way the wind was blowing. You could get rich.

Back in Singapore, Perumal continued his own small operations, publicly listing games between his friends, manipulating the outcomes, running the betting, making a few thousand here and there. But he was searching for bigger game, having gotten a taste for it, higher stakes, greater liquidity in the market. He searched for any usable angle. Bookies would take bets on anything, even friendly matches between company teams. Perumal fixed games between employees of hotels or nightclubs or corporations, graduating a level. These were existing teams, however amateur and marginal. They weren't clubs that he had arranged from thin air. He couldn't control every aspect of the match, as before. He had to concentrate his efforts. He realized that every player didn't need to be in on the fix, just the goalie and the central defenders. He could even get by with just the goalie, if he had to, as long as the goalie reliably allowed the other team to score. Perumal learned that paying the attacking players, or even the midfielders, was throwing away his money. He paid the players to lose, not to score, not to win. As he looked around the field, Perumal watched the odd fan engaged in the action from afar, believing it to be real. The scale did not compare, though the feeling was the same. Perumal experienced the stimulation that Kurusamy must also feel. It was the power to deceive.

Perumal's profits rolled in, but they rolled right back out. The money he earned on his fixes couldn't back the kinds of bets he had to make in order to be taken seriously in the Malaysia Cup. When you bet big and you bet often, as Perumal did, you're bound to lose big, especially when you're not in on the fix. Perumal found himself in the hole for $45,000. He didn't know who held the marker. He had placed the bet through a friend. The friend had "thrown" the bet to a runner, who had thrown it to an agent, at which point the bet had mingled with the thousands of other bets that made the circuit appear tangled and confused. It wasn't confusing to everybody. One person could see through the confusion.

They said that Pal Kurusamy controlled ten of the fourteen teams in the Malaysia Cup, directing the clubs and circulating the players. Himself, he moved around in a big Mercedes.

Pal was tough, unrefined, the richest guy in the game, known to bet millions of dollars on a single match. He didn't mind letting people know that he had made more than $17 million from match-fixing, and this in only five months. Police and politicians depended on his payouts. Criminal groups acknowledged the necessity of his network. For a time, Kurusamy was one of the most powerful people in Malaysia.

Kurusamy punched Perumal in the midsection. "Pay up your bet," he yelled at him. Several of Kurusamy's enforcers had approached Perumal at a local stadium. They brought him to the Boss's place near Yishun Park, in Singapore's Sembawang district. It didn't take long for Perumal to understand that his $45,000 bet had gone all the way up to the Boss. Kurusamy punched him again. Kurusamy was a small man, but Perumal knew better than to fight back.

Kurusamy also knew better than to push too hard, because he was always on the lookout for an edge. He knew that Perumal was fixing. It was his job to know. And a man who was fixing, at any level, might someday become useful.

Perumal wasn't sure what to do. He was prone to looking for an exit route, rather than a solution. But he kept in mind the story of Tan Seet Eng, a Chinese-Singaporean horse-racing bookmaker. Eng, who went by the name Dan Tan, was associated with Kurusamy. Yet even he was forced to flee Singapore when he couldn't cover a large soccer bet, hiding out in Thailand until he could negotiate a payment plan. This was a common story in the world of Singapore's bookies and betting, one that Perumal wanted to avoid. If you were out of Singapore, you were out of the action.

Perumal eventually settled his bet. That was enough for Kurusamy to invite him to his regular poker game. Perumal could hardly keep up, the stakes were so high. Money meant everything to Kurusamy and his circle, although it was clear to Perumal from the action at the poker table that money for them held no value. So much cash was pouring in from Kurusamy's fixing enterprise that he barely had time to account for it. Perumal would sit at Kurusamy's side and watch captivated while the Boss handed out stacks of hundred-dollar bills without counting them, as players, refs, and club officials from Malaysia and Singapore paraded through his office as though he was their paymaster.

Perumal watched and learned how fixing was done at the highest level. How to approach a player in false friendship. The way to pay him far greater than the competition, in order to poach him. How to use women to trap players. How to develop a player, then pull strings to get him transferred to a club under your control. How to threaten someone else in the player's presence, so that he would get the message without feeling in danger himself. How to take a player shopping, buy him some clothes, some shoes, make him feel special, as you would do for your girlfriend. How to follow through on a threat if a player resisted your demands.

Perumal also saw that even a figure as important as Kurusamy still had to bow to the Chinese in gambling circles. The Chinese ultimately held every big ticket. Not only did China have the largest mass of people the world, as well as a rising economy, but it also had the strongest organized crime network in Asia, the Triads. All down the line in the bookmaking business, Chinese controlled everything of worth and importance.

Kurusamy was undeniable, but he was not the only one. Perumal watched teams staying in the same hotel get friendly with one another. Club officials had drinks together in the lounge. One team needed a win to advance in the tournament. The other team had already gained the next round. Money exchanged hands. Or sometimes just the promise of a return favor. It was easy. No victims. It was just the way things were done in Asian soccer. To Perumal, it appeared that everybody was in on the fix, and that nobody was trying to stop it.

He watched players inexplicably miss the net on penalty kicks, and he knew why. The talk was in the market, and if you listened to the talk, you could make some real money. But the money was fleeting. It came and went. Whatever he made fixing, he ended up betting on English Premier League matches, on UEFA (Union of European Football Associations) Champions League matches. Sometimes he won, but usually he lost, because he never knew how those games would turn out. Sometimes he had just enough money to ride the bus down to the stadium, which had become the only place where he hoped to make a few dollars.

In his own fixes, Perumal was learning the valuable lessons of experience. He learned that the fix was not always easy to complete. Players were unreliable. They wouldn't follow directions. They would score when they were supposed to be scored upon. They were sometimes hungover, or they just didn't care. Perumal would watch as the clock wound down on a match, and all he needed was his chosen team to let in one more goal, but sometimes it just wouldn't come. He would harangue the players, but it was clear that even though he paid them money, they didn't feel like they owed him anything. To them, he was just a small-time criminal. He couldn't control them. He was missing something.

Perumal would escape it all at Orchard Towers, Singapore's "Four Floors of Whores," a shopping complex that turned into a sprawling boudoir in the evening. Here there was business to be done. Perumal mingled with soccer players there, many of them foreign players, the high-priced imports with the disposable income that Perumal was trying to secure for himself. As the European players tossed money around and as the girls laughed and wanted in on the action, Perumal sauntered into their circle. He approached one of the players, this time with a new strategy.

Michael Vana was Czech. Perumal knew him from watching Singapore's Geylang United matches. And from what Perumal could surmise, Vana was disinterested. At times, he was the strongest player on the field. At other times, it was hard to pick him out of the lazy back-and-forth of the play. As he spoke with Vana over the music at Orchard Towers, Perumal asked him to win.

Perumal had been fixing single games by compromising the defenders and goalkeeper, compelling them to allow the opposing team to score. Now he saw how the fix could work in another way, with a foreign player who was slumming, on the downside of his career, stuck in an Asian lower league for the nightclubs, the easy money, the women, not the glory that he had once imagined, but which had long faded from his aspirations. In those nights at the Orchard Towers, Perumal realized that the players were just like he was, living without a thought for tomorrow, concerned with money only to spend it. Perumal and Vana locked eyes in agreement over the flashing lights of the action.

Perumal instructed Vana to jog along with the rest of the players throughout a game, until that moment when he needed a goal. Perumal would then shout from the stands, like an impassioned fan. That was the signal, and Vana would exert himself. In the first game that Vana played for Perumal, he scored four goals. Vana easily controlled the intensity with which he played, especially since he was superior to the competition he faced. The partnership thrived. Things went well, so successfully and profitably that Vana started suggesting fixes. Perumal realized that he was not the only one getting addicted to easy money.

Perumal was liquid again, and he rejoined Kurusamy's poker game. He wasn't consistently winning at Pal's table, but he was bragging plenty. The Boss listened closely to what Perumal said, even if he didn't let on. And soon Vana had slipped through Perumal's fingers, gone to work for Kurusamy. Perumal was left with nothing besides a costly lesson in the fix. Players had fleeting loyalty. Fixing partners had none at all. Years later, such realities would upend the high life that Perumal had constructed for himself.

There was another lesson that was more valuable, though Perumal was not ready to learn it. Since Kurusamy had many influential people on his payroll in Malaysia and Singapore, he felt comfortable enough to boast. He had spent ten years in prison, starting in the 1980s, and through that despair had attained wealth and criminal authority. But he became too public. The king of this "victimless" crime hadn't figured on the pride of the victim. Kurusamy wasn't concerned about defrauding bettors or preying on the morality of the players on the field. But he would have profited by understanding that he was lampooning the state. In 1994, Singapore's Corrupt Practices Investigation Bureau (CPIB), the terror of local criminals, initiated a match-fixing crackdown. Kurusamy was arrested.

He wasn't the only one. In September 1994, a Singaporean tournament called the Constituency Cup was coming up, and Perumal phoned a player, proposing a fix in the competition, offering $3,000. In light of the CPIB crackdown, the player reported the approach to police. The authorities researched the phone record. They traced the call to Perumal's residence, and in short order, Perumal had a new residence: prison. But in time, he would soon go further afield than he had ever imagined.

Bail was Singapore's beautiful game, as Perumal and Kurusamy quickly gained their liberty while awaiting sentencing. But their business was shackled by the CPIB. Fixing was too hot in Singapore and Malaysia. The cash had ceased flowing. Kurusamy needed to find another way.

Kurusamy had no other way, no other place. He was uneducated. He spoke only passable English. He was not a man of the world. His world was the Malay Peninsula, with its government and police officials whom the Boss knew by name and shared history. Now this world was off-limits to him. Kurusamy developed an idea. Like the many goods that flowed out of the Singapore port, one of the busiest in the world, export was the key to financial mobility. The Boss summoned Perumal. "Go to Europe," he told him.

Perumal traveled on the passport of a friend, easily slipping off the island, breaking the bounds of his bail agreement. He traveled with a partner. The two flew to the United Kingdom, the center of world soccer, where the inhospitable weather surprised them. They weren't in Asia anymore, and they realized that they had wandered into the deep end of the pool. Back home, they had been suave operators. England neutralized any special powers they thought they possessed. They didn't know any players. They didn't know any cops or politicians. Wandering nearly without aim, they found their way to the training grounds of Birmingham, and of Chelsea, the latter one of the biggest clubs in the game. Like rank amateurs, Perumal and his partner posed as journalists.

This was the land of Ladbrokes and William Hill, a sophisticated, legal gambling market that provided the Englishman with a betting slip to heighten his interest in a match. But this had nothing on the Asian marketplace. Betting in Asia was not for fun, or even for watching a game. It was serious business, the business of cultural addiction, and it was about to grow exponentially, making the English market look like a child's hobby. The Singaporean, Indonesian, and Chinese had no favorite teams, just favorite bets, those that appeared winnable.

These billions of people drawn together in identical behavior constituted an enormous market. Ladbrokes had the name, the veneer of English respectability. It had little more. The market and the power were Asian. But few people knew this. Not yet.

When Perumal approached players in Great Britain, they turned their backs on him, walked right past him, looked right through him. When he did happen to get close enough to players to make his proposal—£60,000 to enhance a match— players laughed at him, then reported him. Word got around, and soon coaches and administrators were running Perumal and his partner off their grounds. Most insulting of all: no one ever called the cops. They didn't take Perumal seriously.

Back in Singapore, Kurusamy was furious with Perumal's lack of results, though there was not much he could do. His trial was approaching. Perumal himself stood trial in January 1995. The court convicted him for match-fixing and leaving the country on a false passport. He was sentenced to one year in prison.

When Perumal received parole, eight months later, little had changed. Match-fixing was still too hot in Singapore. The country's international reputation was at stake. How could Singapore be known for best business practices when its most public events constituted a fraud? The CPIB set out to eradicate fixing in Singapore. The Boss still had to make his money. The United Kingdom had proven impenetrable, for now. But there was another market, and it was even bigger.

Kurusamy arrived in the United States flush with cash, connecting in New York for a flight to Atlanta. Perumal joined him, as did several others from a gathering Asian syndicate. The opportunity before them was worth a concentrated effort.

In Atlanta, they blended in with all the other tourists who were there for the 1996 Olympics. They hung around the hotels, the stadiums, and practice fields where they might encounter players for the sixteen national teams included in the soccer tournament. Olympic soccer was a jerry-rigged competition that FIFA tolerated, so long as it wouldn't infringe on the popularity of the World Cup. After limitations on players' ages and levels of experience diluted the rosters, the result was a marginalized round-robin that hardly did justice to an Olympics competition of the world's most popular game.

All the same, the betting on the Olympics soccer tournament would be worth the effort that Kurusamy and Perumal put into manipulating it. Five different cities in the eastern United States hosted the Olympics soccer competition. Kurusamy and Perumal traveled to one of the venues—Birmingham, Alabama—where the tournament's Group C played in Legion Field. There, according to Perumal, they approached Mexico's goaltender, Jorge Campos, one of the most well-known players in the international game. The Singaporeans attempted to corrupt Campos, but he turned them down.

A partner of Kurusamy, a middle-aged man whom Perumal knew only as "Uncle," had made contact with players from the Tunisian national team before the games started. He claimed to have struck a deal with the team's defenders and goalie. Perumal couldn't be sure, but it appeared to him that Uncle wasn't exaggerating his influence over the Tunisian team, which lost all three of its Olympic matches, 5–1 in aggregate.

It was Perumal's first taste of success while abroad. He saw how it might be done, how you could approach national team players, how willing they were likely to be. And when he returned to Singapore, Perumal wished he had simply stayed in America.

Singaporean police had issued another warrant for his arrest. He spent the night in lockup at the CPIB holding pen. The following day, as officers escorted him to a car on the police grounds, Perumal slipped one of his handcuffs and ran for it. He tried to scale a fence, but it was too high, and the officers dragged him back to the ground. At court, a judge sent him down for two years—plus extra time for attempting to escape custody. Police also picked up Kurusamy. The Boss ended up spending two years in solitary confinement.

When Perumal was released, in 2000, while those around him were excited about the prospects of the new millennium, his hopes were dim. He was thirty-five years old, a convicted felon with no professional skills, no references, and no viable financial prospects. There was only one thing he had ever known how to do. And his time in prison, where he shared a small cell with a dozen men, where a small bowl served as his toilet and his drinking cup, this period had not caused him to develop his talents.

When Perumal consulted the schedule and noticed that a match between two S. League teams was a few days off, he got an idea. There was still too much heat on fixing. Perumal didn't trust the players. They were liable to turn him in, notifying the CPIB to save themselves. But there were other ways to influence the outcome of a match.

The strongest player for the Woodlands Wellington club was a Croatian import, a midfielder named Ivica Raguz. Perumal had watched enough Wellington matches to understand that the team's chances were largely dependent on Raguz's performance. If Raguz happened to miss a game, and if Perumal happened to possess this knowledge before a bookmaker knew that Raguz was going to sit, Perumal stood to gain. Perumal had always considered match-fixing a victimless crime. The only people who got burned were the bookies, and they were dealing in such high volume that, Perumal rationalized, his manner of fraud made little impact. Perumal had developed a philosophy by which he was the patron of soccer's lost souls, the financial champion of players whom the establishment paid less than a living wage. Perumal was the one who augmented their salaries, made their lives possible. Financial desperation may have altered his character. Maybe it was prison. Or maybe the elaborate stories that he was telling himself and others were the convoluted justifications of someone who would indeed do anything to beat the system.

Perumal hired two Bangladeshi men to assault Ivica Raguz before Wellington's next match, against Geylang United. Perumal and a friend placed a bet, 30,000 Singapore dollars, on Wellington's opponent to win. Then they sat back and waited to hear from the Bangladeshis. But the call from the Bangladeshis, confirming that they had incapacitated Raguz, never came. Raguz was a large, heavily muscled man, and when the Bangladeshis saw him in person, they froze. They decided this job wasn't for them. But the bet had already been placed. Perumal had to do something. He had to do the job himself.

With money on the line, and a crude sort of match-fixing proposed, Perumal proved his industry.

He lay in wait in a stand of bushes near the Lower Seletar Reservoir, in northern Singapore, where Wellington's practice field was located. Perumal held a field hockey stick in his hand. He waited for some time before Raguz appeared. When Raguz did materialize, a teammate was by his side. This was Perumal's chance to abandon his plan. He hadn't yet crossed over from fraud to felony assault. His victims showed no visible bruises. He hadn't threatened anyone's health or livelihood. But desperation outweighed all concerns. Perumal didn't want to miss the chance. He stepped out of the bushes, approaching the men from behind. He swung his field hockey stick, striking Raguz on the right knee with the dull, hard wood. Raguz and his teammate managed to escape, but the damage was done. Raguz didn't play against Geylang United, and the bet came off.

Perumal had little time to enjoy his winnings. A few days after the match, police arrested him on assault charges. He spent another year in prison.

When he was released, in September 2001, Perumal kept to himself. He felt like a pariah. He kept away from the few friends who continued to associate with him. Fixing was off-limits, even if he had wanted to revive his connections in that world. But he had to earn a living somehow. Regular work was not an option for Perumal. He had never held a traditional job. With his criminal record, he would not have been able to find honest work that equaled his conception of himself. Instead, he opted for credit card fraud. Match-fixing had earned Perumal fairly light sentences—six months, a year. Considering fixing's financial potential, the risk of such limited prison time seemed worth the gamble. However, institutional financial crime was serious business in Singapore. When the bank that Perumal targeted traced the source of the fraud, it was easy to identify the perpetrator.

Perumal was too poor to hire an attorney, and he instead represented himself at trial. Because of Perumal's previous escape attempts, the judge ordered him to wear handcuffs in court. It was humiliating. And inevitable. Perumal wasn't surprised when he was convicted of fraud. He understood the evidence. He also figured that his days as a match-fixer were over. He had already begun to conceive of how he was going to make a living when he got out of prison. But all of this thinking stopped, as he was staggered to learn of the penalty he would have to pay. When the judge pronounced a judgment of four years, Perumal wanted to hide his face in his hands. But he couldn't, since he was shackled.

CHAPTER 5

Chris Eaton was at the wheel of his Ford Fairmont. The headlights up ahead belonged to a refrigeration truck that was crossing over into his lane. Eaton had little time to react. Instinctively, he swerved to the right. Eaton was driving in the left lane, as is the way in Australia, in a right-hand-drive car. A simple twist of the steering wheel. Debbie, mercifully asleep against her pillow, accepted the force of the collision. Everything transpired at highway speed.

Ian's front teeth were knocked out, and his face was severely lacerated. Eaton's daughter, Sarah, was just fine. At the door of the ambulance, parked across the M31 highway, Eaton looked on. He knew the protocol. He was a cop. He had looked the victim in the eye before, delivered the news. The victim was the last one to hope for a good outcome, and Eaton knew better than to do that now. He watched the paramedics work, and he knew. The whip of the crash had snapped Debbie's brain stem. She was gone. "Life was to be lived for Debbie," Eaton says. "Which was good, since she had such a short life."

Eaton was a widower at twenty-nine, with two young children. He couldn't handle the variable hours of shift work any longer, nor long-term investigations. He had to find stable work. This is how the man who wasn't suited to be a bureaucrat became just that, joining Australia's federal police union, eventually becoming its chief. He thrived in the union, learning how to operate in a political environment, though a rough-and-tumble Australian one, aided by what he had endured. "I developed a healthy cynicism as a result of the tragic parts of my life," he says. "It made me see the realities of life for what they were." Eaton remarried, and with his second wife, Kathie, he had twin daughters. He spent the rest of the 1980s and '90s gaining administrative experience, an Aussie cop for life, it appeared, until an unexpected opportunity arose, expanding his policing portfolio in ways he had never imagined.

In 1999, at the age of forty-seven, Eaton broke from his established career path, challenging himself in a foreign world. He joined Interpol. Headquartered in Lyon, France, Interpol was the eyes and ears of international law enforcement, the second-largest intergovernmental organization in the world, after the United Nations. Interpol didn't arrest suspects. It served a more critical function in the modern, globalized environment. Interpol was the one agency that could serve as a liaison between various national and local police forces, a hub of international criminal intelligence. When a suspect fled one country for another—or, worse, one continent for another—Interpol was instrumental in tracking this person and connecting the relevant policing agencies in order to apprehend him. Interpol was like an international FBI, which made sense to Eaton, from his study of J. Edgar Hoover. As international borders, especially European borders, disappeared, as technology shrank the world, Interpol's role enlarged.

However, when Eaton arrived, Interpol was technically behind the times, and woefully so. The agency still dispatched messages to foreign branches by telex. Among various initial jobs, Eaton managed the implementation of a new communication protocol, Interpol 24/7, which replaced the telex system. He worked as the chief of staff for Interpol's president. All the while, he attended a school of new social and professional manners. He had come from a continent unto itself. Australia in its isolation produced some of the most professional, energetic, and cooperative police officers in the world. But they had limited experience globally. They knew a little about Southeast Asia. But for Eaton, this didn't compare to the astounding complexity of working in Europe, with its sophistication, with its fifty countries and dozens of languages, customs, and legal codes.

At Interpol, Eaton also mixed with African, Middle Eastern, Asian, and American colleagues. A great reader, he now became a great listener, coming to understand what other cultures valued, how they operated. He made quick friendships with his counterparts from Germany, Austria, Russia, Thailand. Eaton further burnished his international credentials when Interpol lent him to the United Nations' independent inquiry committee, which was investigating the Iraqi Oil-for-Food program.

Working under Paul Volcker, the former chairman of the U.S. Federal Reserve, Eaton traced the sources of Saddam Hussein's wealth. Eaton had come a long way from the beat in St. Kilda. He was learning the skills that would transform him into a cop who could capably and imaginatively combat an international criminal conspiracy.

Eaton had come to the attention of Ron Noble, Interpol's secretary general. A tenured professor at the New York University School of Law, Noble had served as an undersecretary at the U.S. Treasury Department before coming to Lyon. He was credited with sweeping structural reform that revitalized Interpol after the terrorist attacks of September 11, 2001. But he was frustrated. Interpol's Command and Coordination Center was brand new, a reaction to the rising global terrorist threat. It was designed to be the room through which all of Interpol's critical information flowed in up-to-the-minute fashion. In practice, the center was underutilized, noncritical, an underperforming asset. Noble recalled Eaton from the UN in order to fix the problem.

Noble knew that Eaton was a talented administrator. Eaton was aggressive. In pressure situations, he acted calmly, assertively, insulating his subordinates from distraction and giving them the assurance to perform. When he returned to Lyon, Eaton went about transforming the command center into the innovation that Noble had envisioned. In short order, the command center became a hive of activity. A massive screen hung on the main wall, and it displayed the status of active incidents and investigations from across the globe. Operators at individual desks communicated to international police agencies in Russian, French, Spanish, Urdu, Arabic. The world of crime and cops generated an intense, unending flow of information, which Eaton's seventy-five charges coordinated with increasing adeptness. A serial killer on the loose in Southeast Asia, a terrorist incident in Africa, a drug suspect arrested in South America, a prison break in the Middle East. As Eaton's daily attention switched from terrorism to organized crime to genocide, he learned the value of sharing live operational information with the people who could utilize it to put an end to the victimization of others.

When this sharing didn't happen, Eaton grew furious, then morose. He watched in disgust as national police agencies greedily hoarded information about a Swiss pedophile who had traveled around Europe on a thirty-year killing spree. This only confirmed Eaton's belief in the need for international cooperation. It didn't matter who got credit for solving a crime or making a collar. All that mattered was getting it right in the end.

Within a few years, Interpol's command center had become the single most important repository of operational data and information in all of international policing. Cops in the field contacted the command center because they believed that Interpol—through Eaton, its manager of operations—would react with the information and assistance that would make a difference in their investigations. Each day, as Eaton scanned the command center's big screen of human frailty, he knew that he was harboring a secret frailty of his own.

Eaton took some getting used to. Colleagues who met him for the first time often found him crass, direct, a little touchy. But once the heat rose in the command center, these same people discovered that there were few senior Interpol officials who were more capable, more fraternal.

Eaton displayed a striking, fundamental dedication to the job. Most people who worked at Interpol were there by appointment, on temporary assignment while still employed by national policing agencies. They were there to make professional contacts, to pad their résumés—there for an education in French wine—until their real bosses called them home. Eaton was an Interpol employee, so he had a stake. But there was something more. "Always remember what you're doing this for," he would routinely tell those in his charge. "What you're trying to do is help the police officer in the field."

One night, Eaton was leaving work, making his way around the command center to shake hands with each person on duty, as was his custom. Word arrived of a South American prison break. When Eaton learned that one of the escapees had shot and killed a cop, he hung up his coat. He slumped in a chair, identifying with the victim.

He phoned one of his underlings, whose expertise he required in order to dispatch the Interpol notices that would aid in the search for the suspects. The employee said that she was home, and that she would come to office once she had finished her dinner. "The only reason you have food on your table is because of these police officers," Eaton told her. "Get your ass in here. Now." Eaton's professional behavior left no doubt that he was operational, not political, and come what may.

When his passion was inflamed, Eaton's voice would boom. His words would come out in a high-vocabulary jumble, and it might be hard to understand him, especially if you weren't a native English speaker, which was true of many at Interpol. Although his arguments were often correct, the bluntness of his debating style derailed him from a path to the top jobs in Interpol's highly political environment.

Often over the years, he and Ron Noble differed. Yet they retained mutual respect. One night over dinner in 2008, Noble told Eaton: "You might be the only person who is more loyal to Interpol than to me."

Eaton "aspired to leadership at Interpol," but he was not obtuse. Such a determined cop, such a disinclined politician. His fundamental political flaw was what made him operationally effective. He was unforgiving. But he was not perfect.

He had had a liaison with a Frenchwoman he had met in Lyon. In secret, there was a daughter. Eaton's marriage to his second wife, Kathie, ended—though, he says, not acrimoniously. The two remain in cordial contact today. "My wife was a good housekeeper," he says. "She kept the house. She deserved it."

Approaching sixty, Eaton's career had stalled, advancement at Interpol closed off to him. His personal life was an open question. But unlike many others of his age, he was not unduly discouraged by the future. He believed he had more to do. He had energy. He was an expert in not only the way that international organized crime operated, but also the way that international police did its business—and how it might cooperate more effectively in combating global conspiracy. Eaton had acquired the knowledge and skills that come to only the adept, energetic, well-placed international policeman. All he needed was a place to apply them.

CHAPTER 6

Hong Kong's Wooloomooloo Steakhouse attracts a busy lunchtime crowd. On the thirty-first floor of the Hennessy building, the restaurant overlooks Victoria Harbor, toward Kowloon and mainland China and all of the money that has transformed global sports betting.

Patrick Jay works his way through a cut of meat. Jay is the head of the sportsbook at the Hong Kong Jockey Club. This may be the most profitable sportsbook in the world, though such rankings are impossible to calculate, given the nature of the business. Jay explains that the Hong Kong Jockey Club handles roughly $6.5 billion in betting on soccer per year. From its entire gambling portfolio, the book takes $1 billion in profit annually. The Hong Kong Jockey Club is the largest taxpayer in Hong Kong, representing 8 percent of the local budget.

Jay is a tall, large-boned man, with the gregarious and happily ravenous manner of someone whose strategic decisions have guided him to a windfall. He projects the attitude of that rare animal, the winning gambler. Jay is one of an expanding cast of Englishmen come east. They carry expertise in the traditional, respected, English way of making a book—at shops like Ladbrokes and William Hill—and they now apply these business principles to Asia, where their experienced hand is welcomed. The Asian market has grown exponentially in the last decade. Jay estimates that the market represented about $100 billion at the turn of the millennium. Today, he says, Asian gamblers wager $1 trillion on sports per year. "The numbers are absolutely unfathomable to everybody," Jay says. "People back in the U.K. don't believe it. If you show them financial numbers, they say, 'You're making this up. You got Enron to do your accounting for you.'" It is not only the size and growth of the Chinese economy that has attracted so many in Western gaming. Nor would adventure be a sufficient motive for someone as oriented to business as Jay to relocate this far from home. It is habit most of all that draws people in Jay's line—Chinese habit, the role that gambling plays in Asian cultures, the well-documented acquaintance with risk. This, as much as Asian economic dynamism, is the guarantor of continued growth in the gambling business. Jay's research tells him that in Hong Kong, locals allocate upward of two and a half times more of their disposable income for gambling than do people in the United Kingdom. "Asia is not the center of the universe," Jay says. "Asia *is* the universe."

Jay's sportsbook is located at, unsurprisingly, a racetrack. It is public, open, legal. And it is categorized in the minority. Throughout nearly all of Asia, the most active gambling continent, gambling is illegal. It is illegal to bet on sports on mainland China, for this activity is antithetical to the precepts of the communist state. The Muslim religion does not permit gambling for Indonesia's 250 million people. This doesn't mean that legal statutes prevent gambling. On the contrary, illegal, unregulated bookies in China, Indonesia, and all across Asia predominate. Jay claims that the illegal betting market is ten times larger than the legal market. Of the $1 trillion total, he says that $900 billion is wagered in the dark, administered by the criminal entities that finance, regulate, and enforce a parallel industry.

At Wooloomooloo, lunch draws to a close, and Patrick Jay readies to make a demonstrative point. "Look around the restaurant," he says. "What do you see?" There are tables full of what appear to be businessmen in the midst of congenial lunch meetings. There are a few romantic couples sharing their little moments. At other tables, friends speak loudly with one another, then laugh. It is the usual steakhouse crowd, but for one missing element. "No booze," Jay says. He's right. Plates of steaks and potatoes cover the tabletops, but no single glass of beer or whiskey accompanies them. "These people don't spend their money on alcohol. They gamble."

China's market reforms of the late twentieth century incited one of the most remarkable periods of localized economic growth that the world has ever experienced. Throughout the 1990s, the Chinese economy grew at a rate of roughly 10 percent per year. In rapid fashion, this swelling generated both great personal wealth for some individuals and general liquidity in Chinese society.

While this miraculous event was unfolding, so was an episode of even greater global significance and revelation. During this period, the Internet was growing from a computer engineer's curiosity into the world's primary means of commerce and communication. At the moment that many millions of Chinese people all of a sudden possessed disposable income, there was a new place to play with it. When these fortunate Chinese considered how they might float their new wealth for the enjoyment and risk that had long been a central part of their culture, they were presented with a growing number of gambling options online.

The emergence of the Internet not only precipitated the growth of online betting sites, but also improved options for the gambler. Before the Internet, the corner-store bookie, such as Ladbrokes, had little incentive to offer its clients competitive odds. It possessed a quasi-monopoly, defined by location and the immobility of the gambler. Internet betting introduced choice to the betting market. A new catalogue of gambling sites began dropping odds and commissions in the competition to attract business.

In China, the new bourgeoisie within this population of 1.3 billion people flooded the Internet gambling market. As the millennium turned, European sportsbooks followed the lead of their Asian counterparts, establishing online portals. Eventually the European and Asian markets began to work in concert, online, following each other's price and line movements, bookies on one continent laying off bets with bookies on the other as part of their risk management strategy. Asian books established European-registered subsidiaries under different names, the client none the wiser. As happened with other industries as they migrated online, in gambling, national borders dissolved. In short order, the Internet enforced global regulation, of a sort, on a largely unregulated, gray-market, underground industry.

"The Asian and European betting markets have come together and created one giant pool," says David Forrest, an economics professor at the University of Salford, in Manchester, England, who specializes in the study of sports gambling. "It's now one huge liquid market. And liquidity is the friend of the fixer. You can put down big bets without notice, and without changing the odds against yourself."

The Internet altered what people bet on, as well as the way that they bet. Twenty years ago, roughly 15 percent of bets on the international sports market were placed on soccer. But as the Internet enabled betting houses to offer continuous propositions based on the various factors of a game in progress—including the time remaining, the score, the players on the field, and the intuition of the bookie adjusting the line and the odds—the rise of in-game betting enhanced the popularity of soccer as a gambling proposition. The game now accounts for roughly 70 percent of the international sports betting market, according to Interpol estimates. The Internet also allowed for a rise in the trading of bets between bettors. International gambling on soccer matches has come to resemble a stock market, with constant fluctuations, numerous propositions, and instantaneous arbitrage.

Along with these changes came heightened scrutiny on soccer matches and the valuable information secreted within them. Patrick Jay tells the story of a grizzled old bookie he worked with at Ladbrokes. "In 1995," the man liked to say, "if the midfielder for Manchester United broke his leg, five people would know about it. His wife, his father, his coach, and his trainer. And me." Now, said the man, if a minor injury afflicts an inconsequential player on an unknown club, "they're betting $10 million on it in that Macau."

Before the Internet, one of the only ways to bet on soccer was on the 1x2 market. The "1" represents a victory by the home team. The "2" represents an away-team win. The "x" represents a draw. The 1x2 market does not incorporate a line, or a point spread. Odds are simply established for the chances of each of the three possible outcomes. The final score is irrelevant. When the favorite builds an insurmountable lead in a match, the gambler doesn't have much incentive to watch anymore. Despite this, the 1x2 market remains the most popular form of soccer betting in Europe.

Almost no one in Asia bets 1x2. The majority of people betting on soccer in the world—and this includes all sizable international match-fixing groups—operate on the Asian handicap and Asian totals markets. Locally known as *hang cheng,* the Asian handicap market takes the draw out of soccer betting. In essence, you bet on one team to win by an assigned handicap, or on the other team to lose by this same handicap. Bookies establish a point spread that recognizes one team as the favorite. They also assign odds to each bet. The odds place a number value on the chances of a proposition and thus the payout on a winning bet. The Asian totals market, on the other hand, offers the chance to bet on the number of goals scored in a match in aggregate, the over-under. Except for a few minor differences, the Asian handicap and the Asian totals markets are identical to the markets for betting on the NFL or NBA, in the United States.

In the European Union, bookmakers have been known to report their clientele to the police for suspicious betting patterns. You must bet with cash or a debit card, not on credit; that puts a hard cap on what you may wager, as you must possess the required funds to cover potential losses. Betting limits are low in Europe. Some European books ban gamblers who win too much or too frequently.

Betting is different in Asia. You can bet anonymously and on credit, and you can win as much as you like. Some bookies in Asia even welcome match-fixers, since the bookies can utilize inside information for themselves, laying off the fixer's bets on other books while adding in some money of their own. These factors contribute to the popularity of the Asian handicap market. This makes it highly liquid. Because there is greater liquidity in the market, bookies take more bets at higher limits.

Most remarkable about the Asian handicap and totals markets is the speed with which they move while a match is in progress. The sportsbooks alter the odds on a bet, and often the line itself, numerous times during a game. In some matches, depending on the play on the field and the betting action that a bookie is receiving, the odds can change every minute, or more often. This affords the fixer a great opportunity, but only if he can think quickly, and only if he has tight control over his compromised players and referees. A match-fixer will routinely sit in the stands at a game, wearing a brightly colored hat, making himself highly visible to his players. He watches his smartphone for the minute-by-minute fluctuations in the Asian handicap and totals markets. When he recognizes the opportune moment to place a bet, he will do so. He'll then take his hat on or off, or turn it to one side or another, communicating an agreed-upon signal to his players on the field. It's time to score, time to let one in, time to instigate a red card, or time just to run out the clock. These movements are often coordinated with the syndicate in Asia, which communicates with the fixer in the stands by phone, relaying its directives through him to the players as the action of the match unfolds.

The general limit for betting an English Premier League match on the Asian handicap market is £50,000, or roughly $80,000. But this limit is "per click," not per match. Each time the bookie changes the line or the odds, setting a new price, the gambler can place a new bet. Bookies routinely alter the odds hundreds of times per match. If a bookie changed the odds just once per minute, a gambler could place ninety bets over the course of a match. That's £4.5 million, or $7.2 million. Each bookie often lists multiple lines per match. There are hundreds of bookies taking bets online. And in Asia, gamblers can bet, through the agenting system, ten times the per-click limit of £50,000. Conceivably, on the Asian market, one gambler could place hundreds of £500,000 bets with one bookie on one match throughout its duration. The size of the marketplace is staggering, as are the nerve and wit necessary to manipulate it.

The new mass and complexity of the global betting market extended a lucrative business prospect to those who could scheme the system in their favor, a new opportunity for crime on a massive, transnational scale. The Chinese criminal groups that were benefiting from the blossoming domestic economy, and which largely controlled Asian betting, were eager to find a way to manipulate the market in their favor.

"We had known for some time that Southeast Asia had a robust illegal gambling market," says Ron Noble, Interpol's secretary general. But even Interpol was caught flat-footed by the millennial changes. "You look at the money and you say, 'Would organized crime like to influence the outcome of these bets?' Yes. The numbers are so huge that nobody can quantify them in a reasonable way."

The windows in Patrick Jay's corner office in the company tower provide a vivid image of the Hong Kong Jockey Club. Situated in Happy Valley, the horse track oval is kept pristinely, lushly hemmed in by the Tai Tam mountains. The valley was once a cemetery, but now it is full of life. Sports facilities populate the infield of the racetrack—tennis courts, soccer fields— and high-rises ring the greater scene. "There's not an apartment here for less than one million dollars," Jay says. As Jay looks down at the track, several coaches run a group of children through drills on one of the infield's soccer fields. Jay says that this is a youth soccer camp.

Then his talk turns away from games, and toward the business lessons he has learned while stationed here so far from home. "In Asia, the world is not as it seems," Jay says. "There is multilayered ownership of banks, sportsbooks, businesses. The Chinese way of doing business is that everybody gets a percentage. Everything you learned at Goldman Sachs, Ford, McKinsey—once you get to mainland China and Hong Kong, just forget it." Jay takes a seat behind his desk. Above him on the wall hangs a Chelsea jersey in a frame. The autographs of the players of the 2009–2010 team cover the fabric.

It is Jay's business to know his competitors in the gambling market, what they are doing, whom they represent, where their money comes from, and where it goes. It is important to know all these things. Yet though Jay sits in their midst, they largely remain a mystery, in the black market and out in the open, their structures raveled across continents and shell companies and empty names.

"The issue in all of this is that we don't know what the issue is," he says. He rattles off the names of some of the largest sportsbooks. "IBC, SBO, 188Bet. There are so many shades of gray. Illegal Asian gambling syndicate? Show me a legal gambling operation. Then show me an illegal gambling operation. Then show me the difference between legal and illegal."

During these recent years of the Asian gambling boom, when soccer's biggest clubs have succeeded in becoming thriving business entities, attaining immense value, a similar blur has overcome the distinction between the game and gaming. Professional soccer players the world over, from the English Premier League on down, play the game with the logo of a bookmaker on their jerseys. Every one of the twenty clubs in the English Premier League has a sponsorship deal with a bookmaker, a club's so-called betting partner. Manchester United's betting partner is 188Bet, one of the largest in the world, registered on the Isle of Man. Sure, this is business—the companies paying for the privilege of association, like any other sponsor—and there is no point in being prim about it. But associations matter.

"These behemoths are bigger than hedge funds," Jay says. "But how much do we really know about them?" He looks again through the window and down below him, at the soccer academy on the territory of the gambling outfit. "That's Manchester United down there," he says. The club is in the midst of an off-season Asian swing, players and coaches running these children through their paces. The most popular club in the world gets around.

CHAPTER 7

Hong Kong has so many bars like this one that it is difficult to remember which names belong to which places. Second-floor slit windows overlook the street. Track lights flicker in the darkness. There are a few girls leaning against the bar in the tired way that professionals lean against bars. It's early, so there is still room to move. Music blares anyway, so William asks for a page out of my notebook. He must write to get his delicate point across.

In Asia's illegal gambling industry, William is what you may call a facilitator, which is why he prefers to utilize an alias. Thus concealed, he is happy to explain the architecture of the industry. The illegal betting world operates on a system of vouching. It is like joining a country club or a fraternity, though with greater ultimate consequences should conditions for the member go south. A prospective gambler gains entry to this world via the recommendation of a contact who already circulates within it. Once he passes muster as someone who will settle his debts, the gambler receives an account with someone who is known as an agent. The gambler places his bet with this agent, who then passes the wager to his superior. This person is known as a master agent, one step up in the hierarchy.

Above this master agent is the super agent. Of these, there are few in the world, and they are almost all Chinese. This man is a trusted, known criminal figure—a face—who can handle upwards of $50 million in bets. The super agent distributes his bets to the many legal and illegal books with which he has relationships, his initial $50 million stake often divvied up into smaller wagers throughout the shadow system. If a gambler loses a bet with his direct agent, someone higher up the chain of command may ultimately come calling.

That is how the money enters the structure. How does it travel through the system? The illegal betting world is so complicated that it requires many diagrams to understand it, and that is why William is drawing. For about fifteen minutes, he constructs an intricate flowchart, representing the way that money circulates through the Asian market. Cash moves from gambler to bookie through Chinese bank accounts in the names of "genuine but anonymous people," as William labels things. Finally, via a "money changer," the lost money of the gambler exits through a corporate entity, into the pocket of the criminal organization. William points to the money changer on his chart. "This is the guy," he says. "Without him, they're dead in the water. This is the guy who takes money from offshore to onshore." Then he says, "You think the chain stops here. It doesn't. It goes up. Way up."

There is so much free-flowing money in sports gambling in

Asia that life within its boundaries could lead only to a single decadent outcome. "This is the best business in the world," William claims. "You stay in the best hotels. Girls are on tap. You drink as much as you want. You eat the best food. Everybody has money. It's a merry-go-round, and it's very hard to get off."

And the center of the spin is Manila. The Philippines is one of the only governments in Asia that permits gambling. Manila's outward-facing sportsbooks accept offshore money only. Chief among these businesses is the book known as IBCbet. Once a gambler establishes an account with IBC, through an agent, he may bet through the company's online portal. Some industry experts estimate that IBC handles up to $150 billion in sports betting each year, a staggering number that may make it the largest quasi-legal sportsbook in the world.

It is quasi-legal because oversight in the Philippines is lax. Ronald Guto, the head of the cyber-crimes division of the Filipino National Bureau of Investigation, says that he has never heard of IBC, though its turnover is equivalent to one-third the size of the Filipino economy. At his humid office in Manila, he scrambles to write down the name of the company on a scrap of paper, and he promises to look into it.

At the seventeenth-floor Manila offices of GWI Business Solutions Inc., it looks like any other day in the life of a medium-sized corporation. GWI provides various services to online gaming companies, including IT and customer support. Beneath it all, here in the RCBC Plaza towers in Makati City, Manila, is where I understood IBCBet resided and it was here that I first sought to meet one of its executives.

It is only later, after months of emails, in a hotel lounge

in Bangkok that is favored by Chinese clients, that I finally meet him. The face is worn. He is a slight man, and he gives the impression that he is taking a break from a weeklong party of endless temptations. His hug is loosely bound, sweaty, enthusiastic, and his eyes are aflame with what he has seen. The man says that he manages IBCbet. He talks about violence and great sums of money. He speaks in loops and parables. "What's impossible is impossible," he says. "What is possible is possible. But sometimes what is possible is impossible." He says that he wants to leave IBC, and his gentleman secretary nods along dutifully, sitting beside him. "I can't believe I got out of Manila alive," says the man. Elsewhere, others have been less fortunate.

Asia may be the king of the marketplace, but it is not the only place where business has been booming. PlanetWin365 is a fast-growing European bookmaker, with $1.2 billion in turnover last year. Registered in Austria, the book enjoys its greatest business in Italy. When PlanetWin365 expanded into the Bulgarian market, the country was new to the European Union. Its government was enduring the growing pains of transitioning to the continental system, reforming its judiciary and political and business practices. Just one European bookmaker had been granted a license locally, and PlanetWin365 experienced great difficulty in establishing itself in Sofia, the Bulgarian capital. Eventually, PlanetWin365 would open more than twenty shops in the country. Business was good, clients attracted to the company's professional approach, which differentiated it from existing local books. "We had better odds," says Giovanni Gentile, the company's spokesman. "It gave other bookmakers problems."

PlanetWin365 was a partner with FIFA's Early Warning System (EWS), which attempts to identify compromised matches through gambling data. As PlanetWin365 made inroads into the Bulgarian market, its executives uncovered an uncomfortable fact. Criminal groups were manipulating a large number of matches in the domestic league. PlanetWin365 provided the Union of European Football Associations (UEFA) with information about fixed matches involving clubs including Zlavia Sofia, Lokomotiv Sofia, Litex Lovech, and Lokomotiv Plovdiv. With the support of FIFA and EWS administrators, PlanetWin365 spoke out publicly against the manipulation it had discovered. In December 2011, officials from the company hosted a press conference at the Sofia Sheraton hotel, where they announced their discoveries. At the end of the press conference, they were told that the president of one of the clubs was in the lobby, waiting for them. "We exited through another door," Gentile says.

PlanetWin365's manager for Bulgaria, Yordan Dinov, spent a considerable amount of time fielding interview requests from local media outlets, rather than running the business that was rapidly expanding. "Our name in Bulgaria was linked to our activity against match-fixing," Gentile says. In April 2012, as the Bulgarian winter began its slide into spring, Dinov traveled to Sofia from his home in the village of Blagoevgrad. He met a man in central Sofia, in the afternoon. They spoke with one another on the street, part of the larger, milling city crowd. The man produced a handgun. He shot Dinov, killing him. Investigators pursued the theory that Dinov's assailant owed him a large sum of money, but PlanetWin365 executives couldn't help but believe that the murder was tied to their efforts on match-fixing.

CHAPTER 8

WORLD CUP, SOUTH AFRICA, 2010

The price of security had gone up. As the 2010 World Cup approached, FIFA officials were increasingly apprehensive that a major terrorist act would befall them. They were also concerned that their players, coaches, and administrators would fall victim to the violent crime that was known to occur at a moment's notice in South Africa, the host nation.

In January 2010, at the African Cup of Nations, terrorists opened fire with machine guns on the Togo team bus as it crossed the border from the Democratic Republic of the Congo to Cabinda, Angola. There were reports that al-Qaeda was planning attacks on the World Cup. Sources claimed that a Saudi terrorist, Abdullah Azzam Saleh Misfar al-Qahtani, was plotting to attack the Danish and Dutch national teams at the tournament.

In response to these unsettling events and reports, FIFA had beefed up security at the Under-17 World Cup in Nigeria, and at the Under-20 World Cup in Egypt, both held earlier in the year. FIFA hired the Freeh Group, led by former FBI director Louis Freeh, to provide field operatives for the competitions. FIFA had also consulted with Interpol. This was an expensive approach, however, totaling $12 million for the two tournaments. The cost of security, in South Africa, for an entire month's worth of soccer, would be staggering.

At the turn of the last century, FIFA had grown out of a European desire to establish a common body to govern international soccer competition. In the one hundred and ten years since then, and especially in the recent decades of the exponential growth of TV rights contracts, attendance, and sponsorship, FIFA had mutated into an insular plutocracy. The organization had developed a culture that outwardly championed egalitarianism on the field of play, while otherwise valuing the right of its leadership to personal enrichment. As it extolled the human virtues of sport, FIFA became an exclusionary territory of a new European business royalty. It was like a social club that withheld access to potential members, standing on the high ground of tradition, until an applicant opened his wallet.

FIFA clears roughly $4 billion for staging the World Cup, but like any organization, it looks for ways to trim its expense budget. Fewer expenses, more profit. But what was the right balance between safety and savings? FIFA officials met with their counterparts at Interpol and the Freeh Group, asking for guidance. Was there a more affordable solution to their security imperative?

Ron Noble deliberated. Noble had promoted Chris Eaton until he couldn't promote him any further. Still, Noble valued what Eaton could provide, and he knew his particular skills. He knocked on Eaton's office door one afternoon in March 2010 at Interpol's Lyon headquarters. He asked Eaton what he knew about soccer. "FIFA needed someone like Chris," Noble says. "He has credibility worldwide for speaking the truth as he sees it and he isn't afraid to confront it. He is a tenacious law enforcement investigator." Noble outlined the opportunity for Eaton. There was a consulting position available at FIFA, coordinating security at the World Cup, in concert with international policing bodies. No matter his vitality, Eaton was approaching retirement age. Noble described the position to Eaton as "a soft landing for your public service."

Eaton had worked in Africa on numerous Interpol assignments. He counted the deputy commissioner of South Africa's federal police as a friend. He was comfortable in Africa. And he liked the sound of the job. Eaton had been looking for a graceful, though stimulating, way to exit Interpol, and the FIFA position appeared to be the right opportunity.

After ten years in Lyon, Eaton left Interpol with the blessing of his superiors, setting up camp in Johannesburg. He visited the ten different stadiums that would host World Cup matches. He represented FIFA in its relations with the various international police and military delegations assigned to protect national teams. He began assessing terrorist threats to the upcoming competition based on information gathered from an array of intelligence sources. His new FIFA bosses were impressed. They had never contracted anyone of Eaton's professionally intense police background. For Eaton, the job was exhilarating, though the work was second nature. He had hit the ground running. However, a new and different threat to the game was about to present itself, and this would knock Eaton off his stride.

In the lead-up to the World Cup, FIFA programmers had developed the EWS computer system, which attempted to identify questionable matches by analyzing betting patterns. A month before the World Cup in South Africa, an EWS administrator received a troubling approach. Peter Limacher, a Swiss attorney, was the head of disciplinary services at UEFA, the European soccer authority. Limacher informed EWS that he possessed intelligence that fixers were preparing to rig several matches at the upcoming World Cup. He said that this information came from a UEFA operative under his direction.

If FIFA execs were stunned, they shouldn't have been. In 2005, as Germany prepared to host the previous World Cup, police revealed the biggest scandal in European soccer. A Germany-based Croatian gambling syndicate had compromised a referee and players in the German second division. (A first-division club was also suspected.) Ante Sapina, a Croat, led the ring. A new, related trial, charging four individuals with thirty-eight counts of fraud, was about to open in Bochum, Germany, on the eve of the World Cup in South Africa.

Marco Villiger, FIFA's lead attorney, oversaw EWS. He wanted to keep these new allegations from UEFA under wraps until he had a chance to probe the claims personally. He phoned

Limacher to arrange a meeting with the UEFA investigator who was the source of the information. The man's name was Robin Boksic.

Through Limacher, Villiger met Boksic, who claimed to possess intelligence reports and documented evidence that several national teams were preparing to fix matches during World Cup competition. Boksic didn't produce the information, but Villiger was so concerned by what he heard that he hired Boksic to work for FIFA in South Africa.

At FIFA headquarters, Villiger informed Jerome Valcke, FIFA's secretary general. Valcke admonished him, explaining that FIFA had just hired a veteran cop, Chris Eaton, to run security at the World Cup. FIFA didn't need two people covering the same ground in South Africa.

FIFA was confused. Historically, the organization's response to match-fixing had been noncommittal, and that wasn't a surprise. FIFA execs did not believe that match-fixing was their concern. Whenever national soccer federations notified FIFA about suspected incidents of match-fixing, FIFA routinely claimed that the responsibility for eradicating fixing from the game lay with the federations themselves. This was a matter for law enforcement. "Every member association is responsible for organizing and supervising football in its country," says FIFA spokesman Wolfgang Resch. "Since FIFA has jurisdiction only over persons affiliated with FIFA, it will never be possible to control parties outside the current system."

FIFA did have a point. It was not a policing agency. It couldn't make arrests. It possessed no powers to investigate. Its business was sports management and promotion. FIFA's only mandated responsibility was staging the World Cup every four years, along with the assorted friendlies that elapsed in the time between the main competition. However, by remaining passive on the issue, rather than playing a strong hand, FIFA nurtured a culture of match-fixing laissez-faire. There was no single body that was more responsible for the welfare of the game. But as Eaton was soon to learn, it didn't appear as though anyone was responsible at all.

Match-fixing was so far down the list of priorities at FIFA that the people who had hired Chris Eaton neglected to mention it to him. Eaton had never heard of match-fixing. In all his years of police work, he had been focused on crimes that made a cop's career, transgressions of apparently more immediate importance—violent crimes, financial crimes, trafficking. Like most cops he did not yet understand that match-fixing, with its hundreds of billions of dollars in play, had become a violation that facilitated the traditional businesses of organized crime. Match-fixing money mingled with drug money, prostitution money, blood money.

Ignorant though he was of match-fixing, Eaton was no stranger to scandal and the damage it could bring to an institution. When he received an email from Marco Villiger, providing a sketch of Boksic's claims, Eaton naturally understood the impact that a fixing scandal could have on his new employer. "This would have sent fear through the hearts of Blatter and Valcke," Eaton says. "They've already got the problem of confounded security at the World Cup, and now bloody allegations of match-fixing?" As Eaton discussed the issue with Villiger further, he grew even more concerned. Villiger had no experience handling controlled informants, so Eaton placed little value on his assessment of Boksic's worth and reliability. Eaton so clearly grasped the threat to FIFA's integrity that he spoke with Villiger bluntly. Eaton would handle Boksic personally, without interference. He didn't want to get entangled in the corporate bureaucracy that he was trying to protect.

Boksic emailed Eaton, using an account under the name of Josip Fejervari, and the two eventually connected by phone. Although his English was halting, layered in a Croat-German accent, Boksic managed to provide what sounded to Eaton like an embroidered account of himself. Boksic claimed that he was an undercover agent for the national intelligence agency of a European country, though he declined to say which one. He was investigating match-fixing. He said that he was passing himself off as a fixer in order to gain proximity to criminals for the purpose of collecting evidence. He claimed that he was massaging two prominent Asian gamblers, one who lived in Norway and one in Malaysia. Boksic said that his intelligence chief had authorized him to cooperate with UEFA and now FIFA, and that he possessed secret transcripts of intercepted telephone conversations. These transcripts, he claimed, proved that there were plans to fix matches at the World Cup for corrupt gambling purposes. Eaton asked for details—which criminal groups were behind the plot, which matches they were targeting—but Boksic balked.

The conversation left Eaton in doubt. He possessed decades of experience working with undercover police, confidential sources, and international investigative operations. It wasn't common practice for an agent who was working undercover to expose himself so readily, especially by phone, and in the modern age, when you had to assume that a third party was listening in to every conversation. In addition, Boksic's evasiveness was also troubling.

In late May 2010, Boksic sent Eaton a text message. Two pre–World Cup friendlies were about to be played in Klagenfurt, Austria: Slovakia versus Cameroon, and Serbia versus New Zealand. Boksic told Eaton that he would attend the games. He claimed that the Serbian national team was planning to fix matches for the World Cup. He wanted to case the team's friendly in Austria, watching for criminal figures mixing with the players around the locker room. On May 26, three days before the Austrian matches, Boksic wrote to Eaton, asking for "appropriate access for all" for the upcoming games.

"Robin, what exactly do you mean by appropriate access?" Eaton wrote in reply.

Boksic responded: "I mean that i can go everywhere in the stadium-cabine of the soccer teams press room etc.al to acces for the games slowakei-kamerun and serbia-neuseeland in klagenfurt 29.05 sat."

"If this is important for what you are doing for us, then of course it can be arranged," Eaton wrote, though his suspicion was evident. "I might come to Austria to be with you for the day."

"Yes it is for cause because we have to know about the serbs and slovaks because people are trying to be the mafia to talk to the players and i want to see what people are talking and there are . . ." Boksic's text trailed off there.

FIFA had not properly vetted Boksic. Yet he appeared to

have a clear pathway to representing the organization in the field in an official capacity, gaining access to international teams days before the start of a World Cup that was now under internal suspicion of being compromised. This was like giving a badge and a gun to a man off the street. The developing Boksic situation was serious enough to Eaton to divert his attention from security matters in Johannesburg. With the start of the World Cup just a few days away, with FIFA gravely concerned about a terrorist attack on the world's largest single event, Eaton boarded a plane for Austria.

Boksic picked up Eaton at the Klagenfurt airport. A full-faced man in his mid-thirties, Boksic had a buzz cut and a little heft to him. He wore jeans and a blue T-shirt, driving toward town behind the wheel of a rented SUV.

Along the drive, Boksic reiterated his suspicions of the Serbian team. He also mentioned Algeria, Ivory Coast, and Nigeria, claiming that these teams also had plans to tank World Cup matches. Eaton listened, feeling him out. Boksic was confident to the point of coarseness. He wasn't tense in Eaton's presence, unless bragging counted for nerves. He mentioned in passing that he had suffered a knife wound while protecting an intelligence services colleague some years ago, an injury so severe that it required six months' convalescence in a hospital. "Six months?" Eaton thought. Boksic said that he often traveled with a security detail. By the time they got to the hotel, Eaton had made his assessment. "I could smell criminal," he says.

The Serbian national team was having breakfast at its hotel, before its match later in the day. Boksic introduced Eaton to Zoran Lakovic, the general secretary of the Serbian team, then proceeded to conduct a conversation with Lakovic in what sounded to Eaton to be Serbian. Eaton puzzled at how Boksic had worked his way into this position, conducting what amounted to a private conversation with a national team official under the guise of representing the world's largest athletic body. He could have been saying anything to the man, and no one from FIFA would have been the wiser.

On the drive to Hypo-Arena, Boksic explained that he had informed the Serbian official that he was representing FIFA. He then warned the Serbian team, because its players had been spotted recently in the company of a well-known Balkan fixer. Eaton bristled. "Assisting FIFA does not mean representing FIFA," he said. "In your future role with FIFA, you are to provide evidence of what you allege to me. FIFA will take any action with players and officials. Not you. You are never to speak with anyone independently again." The men drove on in silence.

Eaton couldn't figure out who Boksic was, or what he was after. Later, at the stadium, Eaton pressed Boksic for hard information. Boksic claimed that he possessed a transcript of a phone call between the Nigerian national team goaltender and a known Singaporean fixer. He said that the Slovenian national team was also under suspicion. When Eaton asked to see documented evidence to substantiate his claims, Boksic motioned for Eaton to look over his shoulder discreetly. A few men huddled nearby. "That's my protection team," he said.

On the ride to the airport, Eaton again pressed Boksic for

documentation. "What evidence do you have about the Serbian players?" he asked.

"My intelligence agency has telephone transcripts of conversations between them and the manipulator," Boksic replied.

Tiring of Boksic's cloak-and-dagger demeanor, Eaton turned to him pointedly. "What agency do you work for?"

Boksic replied, "the BND," the Bundesnachrichtendienst, Germany's foreign intelligence agency. He said that he had worked for the BND for nine years. He claimed that he had recently arranged BND protection for Peter Limacher, the UEFA lawyer who had introduced Boksic to FIFA, because Limacher had received death threats from Croatian organized crime. Boksic went on to say that he possessed contacts in this same criminal underworld, but that he had cultivated these relationships in his role for the BND, clandestinely.

While Eaton was trying to piece together everything that Boksic was telling him, Boksic turned to him. "The BND knows you," he said, in a flattering tone. "They know you're a very good policeman."

Upon landing in Johannesburg, Eaton wrote Villiger, urging him to sever FIFA's association with Boksic. Villiger declined. Eaton then phoned Limacher, at UEFA. Like Villiger, he vouched for Boksic. Eaton asked Limacher what proof he had that Boksic worked for a European intelligence agency. Limacher said that a team of intelligence agents always protected Boksic. Limacher's enthusiasm for Boksic was overwhelming, yet Eaton found the basis for his belief vague.

Instead, Eaton listened to his intuition. It had served him well for so long, tested in hundreds of investigations. He was convinced that Boksic was not a serious informant, that he was lying and exaggerating. If Boksic wasn't who he claimed to be, then who was he? And how had he so easily gained the trust of soccer's two most powerful governing bodies? "These people were living in a fool's paradise," Eaton says.

Eaton had only just begun his affiliation with FIFA. He had no influence, no track record. If the brass wanted Boksic in South Africa, then Eaton would have to work with him. But there was one final protective mechanism that he could employ. He phoned Boksic and asked him to travel to Lyon for a psychological exam. This was standard procedure for an undercover assignment, protection for both the policing body and the agent who would be employing the cover story. Again, Boksic balked, claiming that his BND boss wouldn't permit it.

Competition opened on the nineteenth World Cup on June 11, 2010, with South Africa facing Mexico in Johannesburg. This was the first World Cup ever held in Africa, and along with the vuvuzelas that blared a dull roar across the field of play, this competition brought its own set of distractions.

Eaton spent longs days and nights working alongside Interpol, the FBI, and South African police and military intelligence assessing security threats. Chief among them, as Eaton learned, was al-Shabaab. A Somali Islamic group, al-Shabaab was the deadliest terrorist organization in Africa. Only months before, its leaders had announced that al-Shabaab was a cell group of al-Qaeda. Eaton and his colleagues did their best to keep track of known militants as they crossed Africa's many international borders in the time leading up to the World Cup. Meanwhile, Eaton had tasked his former Interpol subordinates to compile a file on Robin Boksic.

Boksic arrived in Johannesburg soon after the games began. He provided Eaton with a list of six national teams that he claimed had been compromised by fixers, along with the names of a number of players and officials. Boksic reiterated that he possessed documents to bolster his claims, including reports and telephone transcripts. He claimed that his BND boss had now authorized him to hand over paperwork to FIFA, but Eaton waited for the documentation.

He had plenty else to do. Each morning he assessed what petty crime had taken place the night before, and whether any of the incidents posed a greater threat to the tournament. He briefed the head security officers of the stadiums that were to host matches that day. His security responsibilities had so over-whelmed him that he hired a colleague as a consultant.

Fitzroy was a veteran of undercover operations for the Australian Federal Police, embedding with violent drug gangs. On several occasions, he had come near death. Eaton waited to hear what information and impressions Fitzroy would gather from spending time with Boksic.

Fitzroy reported that at times, Boksic was effusive, rambling on through numerous fixing claims, though never providing substantiation. At other times, he was monosyllabic. His mood vacillated. He was vague and defensive. He regularly stepped away from Fitzroy to make or accept phone calls, speaking in German and Croatian. Boksic cultivated an air of danger, claiming that his security detail, German and Israeli in composition, protected him around the clock, in twelve-hour shifts. Occasionally, Boksic claimed, U.S. and Spanish secret service personnel participated. As if to explain the need for such blanketing, he stressed his intimacy with dangerous criminals. He said that organized crime figures had earned millions of euros from the information that he had provided, which he had done to strengthen his cover as "a manipulator and UEFA investigator." This led Fitzroy to conclude that UEFA may have financed its own infiltration.

Eaton spoke with several English journalists covering the World Cup, and he was impressed by their knowledge of match-fixing. They corroborated some of the information that Boksic had ventured. Still, Eaton viewed Boksic much like the South American and European football hooligans whom South African border guards were turning away by the hundreds.

"Boksic states and repeatedly re-states two incompatible cover stories," Fitzroy later wrote in an assessment of "Classic Caress," the name that Eaton had given to the operation. "With the first he maintains he is an under-cover criminal gambler and football match manipulator working secretly for the German Intelligence organisation, the BND. With the second he openly declares himself as an investigator working for UEFA. . . . The logical extension of his own combined cover-stories would require Boksic to portray himself as a corrupt UEFA Investigator to solicit match-fixing information with any credibility. He cannot declare himself to be an investigator one minute, and a criminal the next, to essentially the same client group, without appearing corrupt or simply incompetent."

Boksic continually petitioned Eaton for access to team training grounds, VIP areas, media rooms, and locker rooms for players and officials. Eaton repeatedly denied him. On the tenth day of competition, Boksic gave up, telling Fitzroy that he had determined that the tournament was free of match-fixing. He left South Africa, and FIFA never heard from him again. In late June, Fitzroy sent Boksic two text messages, asking him to make contact. Fitzroy received a message in reply, a question mark.

Eaton and Fitzroy weren't the only ones who questioned Boksic's credibility and UEFA's association with him. "We got an order from UEFA to work with Boksic because he was a guy who could deliver information," says Carsten Koerl, the CEO of Sportradar, a betting-industry watchdog organization. "After two months, we couldn't work with him anymore, because he was telling confusing stories. He was speaking things that didn't make much sense to me."

By the time Spain beat Holland, 1–0, in the final game, winning its first World Cup, Eaton had received a dossier on Boksic. Eaton's contacts in Croatia, Bosnia, Herzegovina, Romania, and Germany painted a detailed picture.

Robin Boksic was an ethnic Croat from Bosnia. During the Balkan conflict in the 1990s, he and his family obtained asylum in Munich. By that time, Eaton's sources reported, he was already a soldier with the Hercegovci mafia, one of the most powerful criminal organizations in the western Balkans. He hooked on with the Munich branch of the Hercegovci mafia, run by Ante Sapina, and by the early 2000s, he was acting as a courier between the two branches of the organization. Boksic's superiors took note of his "high level of arrogance," and they directed him toward more lucrative activities.

Beginning in 2003, informants had reported to Croatian police that Ante Sapina entrusted Boksic with approaching Balkan nationals working as referees, coaches, players, and club managers to discuss what it would take to influence football match results. German police arrested Boksic in 2006, as part of the Bochum dragnet.

Eaton's sources suggested that Boksic was serving clandestinely as a double agent for both the Hercegovci mafia and the German police, providing the latter with information that might harm rivals to his criminal organization's continuing match-fixing enterprise. In the wake of the Bochum case, the Croat leadership of the Hercegovci mafia began to undermine Sapina, attempting to take control of the Munich wing of the organization. Eaton's sources told him that Boksic had provided authorities with information on Sapina, who was arrested again in 2009 for fixing the Croat championship.

It was a murky picture, though clear enough for FIFA's new security chief. FIFA had tasked Eaton with coordinating security efforts at soccer's marquee event, but he had stumbled upon an issue of more immediate concern. Eaton couldn't blame people like Villiger and Limacher. They weren't police. They had no experience in intelligence. All the same, it surprised Eaton to know that a man of Boksic's match-fixing history had walked through the front door of soccer's biggest competition. Had Boksic gone to South Africa in order to fix matches? Eaton thought so.

In Kampala, Uganda, while the World Cup championship match was in progress, two al-Shabaab suicide bombs exploded, killing seventy people who were watching the game on TV. The blasts confirmed the fears of soccer administrators, that an organized criminal group would co-opt the sport for its own ends. And it brought Eaton to his senses, for he recognized a parallel concern. He understood now that match-fixing was as threatening to soccer's integrity as terrorism was to its security. Judging by Boksic's ease of entry, he understood that no one at FIFA was prepared to defend the game. As Eaton pondered this, his mind grasping the critical role that had befallen him, somewhere in the bowels of Johannesburg's Soccer City Stadium, a man brushed past. The man was Tamil Indian, and carrying a fraudulent Singaporean passport.

CHAPTER 9

I asked around, spread the word, 'Which ref is tough enough to do the fix in front of forty thousand fans?'" says Wilson Perumal. "It's not easy. That is a lot of people watching. You have to be really strong to do it. This one guy speaks up, Ibrahim Chaibou, and he says, 'I'll do it.'* It was just before the World Cup, an exhibition. South Africa against Guatemala. He called penalties outside the box. Three hand balls. He did really great. He just took charge and did it. There were forty thousand South Africans in the stadium. The Guatemalans weren't going to say anything." With the Nigerien, FIFA-sanctioned Chaibou refereeing the match, South Africa won, 5–1. The cheering in Peter Mokaba Stadium rang in his ears, and Perumal couldn't help but think of Pal Kurusamy, his mentor, who had once ruled the Malaysia Cup. It was May 31, 2010, and Wilson Perumal was the biggest match-fixer in the world.

---

* Chaibou has denied any involvement in match-fixing. He retired from refereeing in December 2011.

Four years earlier, Perumal had walked out of prison lower than he had ever been. Singaporean authorities had paroled him after three years in custody. But this was no opportunity for rejoicing. Perumal was broke. His fixing network was in tatters. His old contacts had no use for him. The only people he believed he could count on were boyhood friends Dany Jay Prakesh and Jason Jo Lourdes—both Tamils whom Perumal had drawn into the fixing trade.

Perumal had spent his adulthood refining one skill. The last time he had tried his hand at a different manner of crime, the credit card scheme had cost him three years. Perumal didn't have much, but he realized that what he did have had grown increasingly valuable during his prison stretch. The Internet had fundamentally altered the sports betting market. That wasn't the only thing that had accelerated. Perumal now possessed the sense of urgency of one who believes that he has wasted his years and has no time left to dawdle. He started meeting with the connected contacts that he still had, convincing them that he could tap them into the growing gold mine of soccer betting. "He's very charming," Prakesh says. "Wilson can sell you anything." He was selling a concept that any well-attuned businessman didn't need to be sold: globalization.

Not only were Singapore's bookmakers offering limited bets for local league matches, allowing for no liquidity in the market, but Perumal was frightened of the Singaporean justice system. Perumal had a lengthy criminal record now, and he knew that the next misstep he took, however minor, would lead right back to custody. He only had one marketable skill, and it made no sense to apply it in Singapore. If he was going to re-build his match-fixing enterprise, he would have to do it outside the country. He had to take his act abroad.

Malaysian police tailed Perumal through the streets of Petaling Jaya. It was October 2008, the beginning of the rainy season. The Merdeka Tournament, honoring Malaysian independence, was in progress. Officials at the Football Association of Malaysia had never heard of Perumal, but they were eyeing a recent tournament match. The under-twenty national team of Sierra Leone had lost to Malaysia, 4–0. A promotional company had represented Sierra Leone, gaining its national team entrance to the tournament. Perumal was not listed as an officer of the company, yet he spent considerable time around the team, issuing orders to players and coaches. Something didn't add up, least of all the four easy goals that Malaysia had scored. Police uncovered no evidence on Perumal during surveillance, though they remained suspicious of his motive for visiting their modest competition.

In addition to Malaysia and Sierra Leone, the tournament included Afghanistan, Bangladesh, Mozambique, Myanmar, Nepal, and Vietnam. Hardly a World Cup lineup. That didn't matter. Internet betting had changed everything. People from all over the world were gambling on any match that was available on Internet bookmaking sites, and in large numbers. Especially in Asia, gamblers were proving themselves to be voracious to a new degree, playing into the hands of a fixer who knew how to manipulate the system.

But how was Perumal linked with Sierra Leone, a war-torn speck of a country on the far side of Africa, eight thousand miles from Singapore? Financial opportunity knows no distance. Sierra Leone epitomized the new Perumal target. He remembered his foray into European fixing on behalf of Kurusamy, how he clumsily approached players from the biggest, wealthiest clubs in the game. The accounting firm Deloitte recently determined that the average English Premier League player earns roughly $50,000 per week, which is more than the average annual British salary. For a fix, a player might earn a bit more. But it was folly to think that these players would risk their lucrative careers on a fix, especially with someone they had only just met in the parking lot outside the practice field. Perumal had gone about it all wrong, and he wouldn't make that mistake again.

In the late 1990s, approaching players from the top clubs made some financial sense, if not operational logic. The biggest games were the only ones that generated the sort of liquidity in the betting marketplace that allowed a fixing syndicate to place wagers that wouldn't upset the market. But the market had since grown to the point that you could place bets of several hundred thousand dollars on Sierra Leone's under-twenty team without raising alarm. It took Wilson Perumal, who had spent his adulthood in fixing, a short time to read the warping market and develop a new strategy.

No longer was his goal to corrupt individual players or referees. He would corrupt entire national soccer federations. Perumal established a front company, Football 4U International, and he began to travel the world, representing himself as a promoter who specialized in arranging exhibition matches, or friendlies, between national teams. "It was a lot easier when you approach people when you are a company," Perumal explains. "People want to know where you're from. If you're not registered somewhere, people are going to be skeptical of you." Singapore's reputation for conducting clean, corruption-free business made the country's passport one of the most accepted in the world. And it enabled Perumal and his lieutenants, Prakesh and Lourdes among them, to walk in through the front door of federations, particularly ones that had fallen on hard times.

These organizations weren't hard to identify. Perumal signed contracts with not only Sierra Leone, but also Bolivia, El Salvador, Zimbabwe, South Africa, and others, paying them as much as $100,000 in return for the right to select the officials who would referee the international friendlies that he would arrange. Perumal would also handle travel, accommodation, TV rights, and sponsorship, insinuating himself into the federations at multiple levels. The struggling federations soon found themselves suddenly wealthier, and playing against high-profile clubs in the kinds of matches that would draw liquidity to the international betting market, if also a curious amount of red cards and penalty kicks. Perumal knew what he was doing. So did many federation officials, who either turned a blind eye or became actively involved. Meanwhile, Perumal built relations with not only refs and administrators, but players, coaches, and anyone else who might be able to influence the outcome of a match.

It was easy. Once Football 4U had forged an agreement with a national federation, together they targeted an unwitting opponent, then paid a FIFA match agent to sanction a friendly or a series of friendlies. At that point, bookmakers would list the matches, and Perumal was in business.

Perumal's communications with officials from the Zimbabwe Football Association (ZIFA) revealed his opportunistic zeal. Perumal claimed to have found a willing partner in Henrietta Rushwaya, the CEO of ZIFA, and that the two discussed numerous opportunities, including Zimbabwe's October World Cup qualifier match against neighboring Namibia.* By the middle of 2008, Perumal and Rushwaya were negotiating a friendly between Zimbabwe and Iran. In advance of the CAF Champions League, Africa's equivalent of the UEFA Champions League, their communications intensified. On August 17 and August 30, 2008, the Dynamos Football Club, from the Zimbabwean capital of Harare, would face Cairo's Al Ahly, the biggest club in Egypt. Writing from the account of his then girlfriend, Aisha Iqbal, Perumal sent Rushwaya the following email, on August 4:

Subject: Re: Friendly international with Iran

Hello, Maam

How are you doing. Please keep me posted on the Dinamo's [sic] game with Al Ahli [sic]. We want 2 goals in each half and you can get 1 goal after conceding the 4th goal. Reward will be 100,000 US dollars. You will have to take your cut from this sum. CAF champions

---

* Rushwaya has denied any wrongdoing. ZIFA fired her but she has never been convicted of any charges arising out of Perumal's account of their dealings.

league is not very popular yet and that is the reason why it will not fetch more than 100,000.

I can influence the match against Namibia in your favour leave that to me. Please ensure your team gets adequate match practice to ensure a safe passage into the next round. There is close to 500,000 US dollars if your team can go to the next round. Please arrange friendly matches with european nations even if it falls in 2009. airfare will be taken care of.

I have some youth tournaments comming up in november. Do you have a U21 boys team. Please assemble 1 if you dont.

Thank you.

Raj

Please keep me posted on the Iran match.

In its essence, Perumal's new strategy was no different from his first fixes back in Singapore's Jalan Besar Stadium. The difference was the amount of money he was handling. He was now in the big leagues.

CHAPTER 10

SINGAPORE, GOODWOOD PARK HOTEL, 2007

Dan Tan knew about Wilson Perumal for all the wrong reasons. The horse-racing bookie who had to leave town for a while, Dan Tan occasionally worked for Kurasamy's top lieutenant. In the '90s, Dan Tan had heard plenty in Kurusamy's company about how Perumal didn't pay his debts. But Dan Tan understood the changing face of the fixing business better than almost anyone else, and there were few people, like Perumal, who were in positions to capitalize on the new opportunities.

Dan Tan had been involved in some of the first efforts by Asian fixers to try to conquer Europe. While these attempts were unsuccessful, by 2005, Dan Tan had returned to Europe, as investigators in the Bochum trial would maintain, eventually fixing in Italy. It was a testament to the Singaporean's fixing competence and reliability that Italian organized crime found it profitable to partner with them. Ironically, it was Singapore's honest business reputation that attracted organized crime groups in Bosnia, Bulgaria, Croatia, Hungary, Italy, and Slovenia.

These parties organized into a group of shareholders, pooling their finances and logistical resources. The syndicate used money carriers from Panama, Nicaragua, and Slovenia to traffic the cash needed to pay compromised players, refs, coaches, and administrators. Investigators later identified Dan Tan's and Perumal's involvement in first- and second-division matches in Austria, Germany, Hungary, Switzerland, and elsewhere on the continent. The shareholders eventually bought pieces of European clubs. German police subsequently identified the syndicate's involvement in a qualifier for the 2010 World Cup, between Finland and Liechtenstein, as well as a 2009 Champions League match between Liverpool and the Hungarian club Debrecen.

Dan Tan had manipulated matches, but he was more financier than fixer. He grasped the changing landscape in a way that no simple financial backer could. When Perumal got out of prison and embarked on his first international successes, Dan Tan recognized someone who might be good for business, no matter what had transpired between him and Kurusamy. Dan Tan was a man of business, not of emotion. Through a Malaysian syndicate runner they knew in common, Dan Tan summoned Perumal to a meeting.

As Perumal recounts it, the two met in the lobby of the Goodwood Park Hotel, a colonial-era property on Scotts Road in Singapore. Dan Tan had boyish looks, his hair parted down the middle. He gave the impression of an earnest, aspiring businessman, rather than a sophisticated criminal. Perumal was impressed.

"I've known about you for a very long time," Dan Tan said.

"I've also heard of you," Perumal replied.

"You have crossed a lot of people," said Dan Tan. "Never mind. We'll put the past behind us. We'll see what we can do together."

Dan Tan financed Perumal for a few small-scale fixes, as the two felt each other out. Perumal then went to Syria with Dan Tan's support. Players in the Syrian domestic league earned about $1,000 per month. Perumal offered several players $10,000 for a fix. The players proved more willing than capable, and three planned fixes failed. Dan Tan lost money on the operation, yet he continued to work with Perumal, their first big score a friendly between Bahrain and Zimbabwe. Dan Tan and Perumal, with their various skills and ever-widening networks of contacts inside and outside the game, would go on to undermine the integrity of the sport on a massive scale. The syndicate provided all the benefits of cooperation, but this kind of cooperation brought its own risks. "If someone betrays the group or gains benefit only to himself at the expense of others, the other members of the group may cause really serious trouble to this shareholder," Perumal would explain later. "By this, I mean they may put your life in danger."

CHAPTER 11

A new opportunity was about to arise, one that could make Wilson Perumal wealthier than he had ever imagined. Through Dan Tan's mainland Chinese criminal connections, the syndicate had come upon several new betting services. These were Triad-controlled groups that utilized sweatshops across Southeast Asia. Rows of workers sat in front of computers placing $3,000 bets as fast as their fingers could press the keys, the wagers small enough and spread widely enough on enough credit cards to avoid detection by bookmakers both legal and otherwise. Whereas before, the most the syndicate could wager on a match without raising attention was roughly $1 million, depending on which teams were playing, now the total bet could top $5 million. The Singapore syndicate, through Dan Tan's contacts, afforded its European criminal partners access to this new opportunity, which only elevated the volume of business all of the shareholders could do together. If only the syndicate could hold, the potential for profit was boundless.

The players whom Perumal targeted were so receptive to an approach that the first question they would ask him was "How much?" Soon players, referees, and club officials from distant pockets of the world were doing likewise, calling Perumal and his lieutenants directly, their desire for side money was so acute. "Hey, we have a game next week," they would say. "Let's fix it."

One of Perumal's subordinates attended an international youth tournament in Kenya. At the opening of the competition, dignitaries entered the stadium in a solemn procession. Once they spotted Perumal's representative in the stands, they rerouted the procession, climbing the many stairs to greet the man, who placed one-hundred-dollar bills in the hand of each supplicant dignitary.

There was more business than anyone could handle. The fix was on at every level, from players to coaches to refs to federation officials. National coaches would call them from the sidelines during games, telling them, "We're about to let in a goal for you." The head of one national federation called one of Perumal's subordinates while in London. "Oh, I've lost my wallet, and I'm here with my family," the man said. "Could you lend me five thousand dollars?" The syndicate used imposter refs. Not that they needed to. It wasn't long before Perumal had more than ten FIFA refs on his payroll. At the African Cup of Nations, one of his associates claims, he fixed the entire tournament.

A few people involved in the fixing business had become international professional gamblers. Others gave themselves over to debauchery. Perumal was always after business, selling players on the advantages of being in his stable, entertaining them all over the world. He routinely took players to a brothel. The price at one of his haunts was $150. But not for the African player that Perumal once brought there. "The African guy comes out once he's finished," says Perumal. "The guy who owns the place tells me $450. I say, 'What? It's $150.' The guy says, 'No, he did it three times.'"

By 2009, Dan Tan had become Perumal's regular financier. Hidden beneath was the match-fixing syndicate in Singapore, which, according to FIFA, was run by Tan and three other bosses whose legitimate businesses enabled them to fund the payouts and travel expenses that Perumal and his lieutenants required to enact the fix. These men worked in concert with one another, often using the international Hawala system, a sort of organized crime financial honor system, to move sums around the world without detection.

This financial backing put Perumal over the top, and it made him boastful. He once claimed that he was "in better control of the Syrian Football League than Assad was over his people." And he found moral grounding in his activities. He styled himself as soccer's Robin Hood, his payments to compromised players in Africa, Central America, and the Middle East providing funding to buy homes, to send children to school, to secure medical treatment for old and ailing parents. "I would send these kids home from a friendly with ten thousand dollars in their pockets," he says. "Do you know what that kind of money means to a young guy like that?"

As close as Perumal and Dan Tan had become, fixing was

an entrepreneurial marketplace. Perumal was free to sell his proposed fixes to whoever wanted to back them. A Chinese financier offered him $100,000, and with this Perumal returned to Syria. When Dan Tan learned that all was going well in Syria, where he had lost considerable sums with Perumal in the past, he bristled.

Perumal was in Kuala Lumpur in August 2009, where he was overseeing a fix between Kenya and Malaysia, a scoreless draw. He was sleeping in his hotel bed when the door to his room burst open and a handful of men entered. Perumal had used a travel agency that Dan Tan owned, and he believed that Dan Tan had given his whereabouts to the man who stood over his bed. "You're here for the money?" Perumal asked.

Pal Kurusamy replied: "I'm here for the money."

Kurusamy still held Perumal accountable for an $80,000 misunderstanding from the 1996 Atlanta Olympics. Thirteen years later, he had come to collect. Perumal had no choice but to settle the debt. Perumal and Dan Tan would continue to do business with one another—there weren't many people capable of carrying out their elaborate business—but any trust that had existed between them was gone forever.

CHAPTER 12

FIFA HEADQUARTERS, ZURICH, 2010

Qatar was a tiny country of fewer than two million people. But its position along the hydrocarbon-rich Persian Gulf gave Qatar access to the world's third-largest natural gas reserves. It had the richest per capita gross domestic product in the global economy, more than $100,000. Because of this liquidity, the leaders of this small nation believed that they possessed the stature to host the World Cup. With desert heat and humidity beyond human endurance in June and July, the traditional staging time of the World Cup, Qatar, it would have seemed, hardly stood a chance. No matter. Officials with the country's soccer federation were doing all they could do in order to prove that they were honorable members of what Blatter persisted in calling "the football family." By the time Eaton started his job at FIFA, Qatar's bid for the 2022 World Cup had made it to the final round of consideration.

Sensitive to any hint of impropriety at a moment of such opportunity, officials at the Qatar Football Association were interested to receive a provocative email on September 12, 2010, just five days after a questionable match between Togo and Bahrain. The letter, titled "4 NATION INTERNATIONAL U23 YOUTH TOURNAMENT," read:

> *Dear Sir,*
>
> *We are keen to organize a u 23 youth tournament in Qatar from 10 o 14 October 2010.*
>
> *We are prepared to incur the airfare and accommodation cost for 3 visiting teams.*
>
> *We will [be] grateful in [case] you can give us an appointment as soon as possible to discuss further on this matter.*
>
> *Looking forward to your positive reply.*

Qatari soccer officials weren't sure what to make of the email. It was rare to receive such a proposal from an unknown party. The sender of the email appeared already to have contracted three other teams for a tournament. And there was little time to arrange everything—less than a month—in a region, the Middle East, where the professional culture was not known for urgency. International soccer was a politicized world of intrigue and double-dealing, with the World Cup worth perhaps $10 billion to the host economy. On the eve of voting for the 2022 World Cup, was this letter a ruse, meant to undermine

Qatar's bid? The Qataris thought it best to investigate. And Wilson Perumal had made that easy, since he had signed his name at the close of the email.

Perumal's brazenness (he had sent the letter from his own personal email account) evidenced the fact that no one was pursuing him. He had nothing to fear. His email to the Qataris bore the markings of a typical Perumal fix: an overture to an out-of-the-way country, an offer to arrange a series of international friendlies, a collapsed window of opportunity. This time, however, Perumal reached too far. The years of his fixing success had eroded his discipline. He made one critical mistake. He abandoned his well-constructed strategy of approaching poor soccer federations, those likely to look the other way in exchange for the financial assistance that would rescue them from insolvency. Qatar may have had a low international profile, but it was no poor nation with an impoverished football federation. Money was the one thing that Qatar didn't need. Instead, it wanted respectability.

One week after receiving Perumal's email, Saud al-Mohannadi, the general secretary of the Qatar Football Association, drafted a letter to Mohamed bin Hammam, the president of the Asian Football Confederation (AFC). Bin Hammam was one of the most senior administrators in international soccer. Himself a Qatari, bin Hammam was a fifteen-year member of the FIFA Executive Committee, which was meeting underground in the Zurich headquarters to determine which country would host the 2022 World Cup, an announcement that would come in less than two months. Citing the Togo-Bahrain match, Al-Mohannadi's letter read:

We would like to bring to your attention that the agent in question for this match is very much active and in contact with many associations and please find attached the recent email we have received from him with an offer for a friendly tournament.

We recommend that AFC can warn all member associations to be cautious when dealing with the alleged agent.

Bin Hammam passed along this letter to Jerome Valcke. In late October 2010, the letter wended its way to Eaton. "It rattled around the office," he says. "Now that FIFA had a head of security, they had someone to give the letter to."

After the World Cup in South Africa had finished without a serious breach, FIFA hired Eaton as its first head of security. He arrived at FIFA's Zurich headquarters, a $200 million steel and glass structure finished four years earlier. Two-thirds of the building was located underground. The design reflected FIFA's bunker mentality. The organization was losing public confidence under increasing pressure over match-fixing and internal corruption.

Eaton's desk was situated in the basement. Under Swiss law, employers were required to place their charges in sight of natural light. Eaton's office was lit by a shaft of daylight from the building's central atrium. Jerome Valcke, FIFA's secretary general, apologized to Eaton when he showed him to his new workspace. FIFA had been growing rapidly; there was just nothing else available.

Eaton didn't much care. He had plenty to occupy him. When he read Perumal's email to the Qataris, the name at the bottom of it meant nothing to him. Wilson Perumal was one of thousands of suspected criminals he had encountered in his police career. The only difference was that now Eaton was no longer a cop. He had no badge, no gun, no power to arrest. He knew nothing about how match-fixing was accomplished, nor the identity of the influential actors in the market. Eaton was unequipped and uninformed, a crusader in the raw. Curiosity and enthusiasm were all he had.

He continued to receive evidence, including a letter of complaint from several Zimbabwean referees, which named Perumal. Eaton researched the company Football 4U. Another promotional company, Exclusive Sports, had been registered under Perumal's name in Singapore. Eaton learned that it now belonged to a taxi driver, who couldn't say why or for how much Perumal had sold it to him.

In his research, Eaton encountered various names and combinations of names connected to these front companies— Anthony Raj Santia, Blapp Johnsen, Wilson Raj, Hedy Larsen, Perumal Raj—and he struggled to bring order to it all. He found evidence in contracts and in press reports of Perumal's front companies operating in South America, Central America, Europe, and the Middle East. In Africa, players and administrators referred to a match promoter named, simply, Raj. "These things were all happening in rapid fire," Eaton says. "Perumal kept cropping up. The picture became clear quickly." Eaton was learning Perumal's mode of operating. His companies generally paid a federation fee of $30,000 per match, plus expenses of $16,000. Match contracts invariably included no income stream for the company. They simply authorized Perumal to select referees. All payments were in cash. As Eaton pieced things together, his focus gradually shifted away from Robin Boksic. The Singaporean angle appeared even more insidious than the Croatian who had nearly infiltrated FIFA at the World Cup. Realizing how widespread Perumal's activities were, Eaton looked at a map of the world and thought to himself, "How does a copper do an investigation on this?" He needed help.

Eaton petitioned Valcke: he wanted to hire a team of investigators. Even though he had hired Eaton to manage operational security, and not to investigate, Valcke approved the request. FIFA was now showing signs that it intended to play an active role—perhaps *the* most active role—in combating fixing. Over the fall of 2010, Eaton assembled a team of investigators, former cops like himself, choosing his subordinates for their "high energy and high criminal hatred," their affability, their ability to cultivate human sources. One Australian, an Englishman, and a Spaniard. He gave them free rein to operate in a "decentralized, bloody active way," stationing them around the globe—Kuala Lumpur, Colombia, and the United Kingdom. Eaton himself maintained his desk at FIFA headquarters in Zurich, though his passport pages began to fill with the stamps of the world as he coordinated the effort to map the shadow landscape of global match-fixing. "FIFA was supposed to be a nice ending to my career," Eaton says. "But I was busier than I had ever been before."

Time and again through the last quarter of 2010, Eaton returned to evidence that appeared to incriminate Wilson Perumal. Besides Qatar and Zimbabwe, information pointed to

El Salvador, Peru, and Costa Rica. "Perumal had left his footprints all over the globe," Eaton says. "All this was retrospective. We were chasing the guy's tail. But the times between our arrival in a country and his departure were getting shorter."

Looking more closely at the Americas, Eaton uncovered the involvement of Perumal's companies in suspicious matches in Argentina, Bolivia, and Panama. Eaton believed that Perumal's front company had infiltrated the Salvadoran Football Federation. Visiting San Salvador, along with his Latin America operative, Eaton confronted officials at the federation offices, one of whom, a former federation president, demanded six thousand dollars to talk. Eaton wasn't there to pay, or to make friends. He possessed evidence that the Salvadoran national team had fixed, among other matches, a 2010 friendly against the U.S. team, a 2–1 loss in Tampa, Florida. This was the beginning of an investigation that would take more than two years to complete.

The location of Eaton's FIFA office ultimately didn't matter, since he was rarely in it. The only way he was going to find Perumal, or Raj, or Wilson, or whoever this gathering phantom happened to be, was by doing it in the field. As Eaton stepped out of first class and into a balmy black sedan in another equatorial country, he would think to himself or even mutter it aloud: "the thrill of the chase." To gather his prey in the present, Eaton was forced to go back in time, re-creating Perumal's previous activities.

CHAPTER 13

SINGAPORE, 2012

Out of the pedestrian horde of a Singaporean shopping district, Dany Jay Prakesh appears, saying, "I think we have an eye for each other." He takes a seat in a café and calmly orders an espresso. He wears a red Air Jordan T-shirt and a diamond stud in each earlobe. He is in his late forties, and he is in noticeably good shape for his age. There is not a line on his face. One of Eaton's FIFA operatives once explained how Prakesh had eluded their surveillance following a meeting one afternoon. Prakesh sprinted through a hotel lobby, ducked down a service corridor, leapt over a retaining wall, and disappeared on the street.

Prakesh is friendly and engaging, an easy talker, as every match-fixer must be. He brags about the number of goals he has arranged in a single match. He says that fixing is easy.

"Anywhere there is a Central American player, or an African player—even in Europe—it's done." He talks about his old friend Wilson Perumal and how they fell out.

Perumal was riding high, crisscrossing the globe, playing by his own rules, manipulating financial backers for his own gain. He slipped with impunity in and out of stadium locker rooms, nightclubbed with players, met with dignitaries who wanted to be his friend for pocket change. Perumal was an entrepreneur. He was also an operative for Chinese Triads, the Italian mafia, and Russian organized crime, the point man for a fraudulent, transnationally interconnected enterprise worth possibly billions of dollars per year. He would do anything to force the fix. With the boundless cash of the great Chinese economic boom at his disposal, Perumal was everybody's best friend, the one who made it easy to be vulnerable. He was the most high-profile fixer in the world. But this is not a profession that favors publicity.

Perumal's merry-go-round stopped in Costa Rica. He had his hooks into the national soccer federation. He had a day to kill. He holed up in his hotel room and started betting. It wasn't enough for Perumal to bet on games that he fixed. For in this, there was no gamble. It was like playing solitaire. The man who specialized in guarantees found no thrill in the fix. He laid bets on the Chicago Bulls, on Manchester United, on clubs and contests he knew were not fixed. The same instinct that served him well in fixing a match—reading players, feeling the nuance of the action—benefited him when he wagered his money on the unknown outcomes of random games. "He was in that hotel room betting for twelve hours," says Prakesh, "and he won one million dollars."

A windfall is only a distraction from the fundamental reality that you are willing to risk it all. Perumal had been gambling so actively and losing so heavily—"he lost ten million dollars in three months," says Prakesh—that he began engaging in a dangerous game.

The pressure was getting to Perumal when he walked out of Singapore's Changi Airport in May 2009 to find a security guard writing a ticket on his illegally parked Honda. "I gambled three hundred thousand dollars on one game," he thought. "What is a ticket for two hundred dollars?" Money wasn't Perumal's problem. His problem was authority.

Perumal pleaded with the security guard, but the man ignored him, continuing to complete the traffic slip. When Perumal opened the driver's-side door, the security guard slapped his wrist. "You're not supposed to drive away," said the man. "I've called the police." Perumal seethed. He had played tough with players, referees, coaches—people who were easily frightened by someone they perceived as a connected criminal—and it usually worked. But when he turned on the security guard, the man stood firm. Perumal felt the blood rushing to his head. He walked back inside the airport to cool down. With all that he stood to lose, with his criminal past and his current criminal enterprise, Perumal would have been wise to take the traffic ticket and pay it. Instead, he returned to the Honda and attempted once again to open the door. The security guard rushed at him. Perumal stood his ground. The guard bounced off him. Prosecutors later determined that Perumal had used criminal force. He sped off the airport lot, the Honda's front corner panel grazing the security guard's leg.

Police arrested him on charges of assault. Perumal was convicted. Making special note of Perumal's criminal record, the presiding judge handed down a crushing prison sentence: five years. Perumal was devastated. For all of the hundreds of crimes he had committed, he had trouble accepting that this minor altercation with a security guard would cost him the most time of all. Perumal was forty-four years old. He had only just resurrected his fixing career. He couldn't foresee what kind of man he would be after five more years of confinement. Singapore had spawned the world's most prolific match-fixers, who roamed the world in their crimes, and the state did little about it. But Perumal's checkered past had awoken Singapore's judicial zeal. He thought about those five years, and he was determined not to serve them.

Post-sentencing, temporarily free on bail, Perumal frantically searched for a solution to his predicament. Where was the loophole? He came up with an idea. He managed to get a replacement passport in the name of a friend, Raja Morgan Chelliah, but with Perumal's picture on it. If he could make it out of Singapore, he foresaw no problems. Even if the photo of the real Chelliah appeared on a computer screen of a customs official at a European border, he would be in the clear. "White people can't tell one Indian from another," he thought. At home, there was another problem, though easily solved. Singapore passports include the holder's biometric print. Airport immigration officers were the only ones who checked for it. The checkpoints along the roads leading to Malaysia were not yet equipped with the optical imaging equipment that could read the passport thumbprints. He would travel by road.

Perumal had run before, and he had paid for it. He could only imagine what the penalty would be if police caught him this time. And if he did make it over the border, there would be another price to pay. No matter Perumal's hostility for its judicial system, Singapore was his home. His base of operations. His financial backing came from Singapore. He weighed his options, and found that he had only one. As he drove across the Johor Causeway and into Malaysia, he was again a free man, though now a fugitive. It was difficult to say what weighed more heavily on Perumal's mind: his mounting debts or this new priority, to avoid ever returning to Singapore again.

Perumal needed to find a new home base. He moved to London, renting a one-bedroom apartment near Wembley Stadium, paying six months in advance. In the year before the World Cup in South Africa, Perumal left his footprints across the international game. He had players, coaches, refs, national soccer federation officials, and political leaders in on fixes. He fixed league games, high-profile friendlies, youth games, untelevised matches played on patchy grass with a dozen people in attendance—whatever the Asian bookies offered, whatever was bound to initiate in-game betting. Perumal worked in more than sixty countries. No one was tracking him and his crew of runners and bagmen. There was no legislation enacted to prosecute match-fixing. And it was difficult to identify just who these match-fixers were anyway. European police received a tip about a fixer named Ah Blur, which was one of Dan Tan's aliases. When the authorities managed to locate a photo of him, it was fittingly out of focus.

Even when cops happened into a discovery, they didn't recognize it for what it was. Border police at Toncontin International Airport, in Tegucigalpa, Honduras, stopped Jason Jo Lourdes and discovered $85,000 in his suitcase. He was traveling with the goalie for the Guatemalan team. Lourdes said that he was in the textile business, and that he was a bit of a gambler, hence his need for all the cash. On another occasion, border guards found $50,000 in his suitcase. Both times, Lourdes walked.

The halfhearted search for match-fixers recalled the days before the FBI, before cooperation between neighboring police forces, when a criminal was home free the moment he crossed a state line. With no police pursuit, Perumal grew brazen. Bookies routinely employ spotters to relay information from untelevised matches as they happen. Perumal simply compromised several spotters, who then tricked bookmakers with regular updates from games that weren't happening. Nevertheless, these "ghost matches" were listed on international sportsbooks. Underscoring the fact that no one was monitoring their activities, Perumal and his partners created dubious names for the officials of their front companies—Hedy Larsen and Blapp Johnsen—the latter a reference, no doubt, to FIFA president Sepp Blatter. Perumal was laughing at the soccer establishment, and the soccer establishment didn't get the joke.

As Perumal bribed and lied and hoped each time for the

fix to come through, the vagaries of the players and the field, and the knowledge of the sums wagered by the syndicates back in Asia, made fixing a nail-biting affair. He would use a hat to signal the players during matches. If he turned his hat sideways, for example, it might mean that he wanted them to let in three goals in the first half. Perumal took advantage of the fact that FIFA sanctions all matches, but that many matches go off without sanctioning. The only thing that mattered was that the major legal and illegal books picked up a match, creating a betting market for it.

Perumal was interested in those who had no money at all, his search for impoverished soccer federations leading him and his lieutenants to the world's poorest places. "One time, I was led out into sugarcane fields in Trinidad with drug kingpins with guns," Prakesh recalls. "I told my wife, 'I don't think I'm coming home this time.'" Mexico was the lone country they didn't operate in, for the cartels there didn't tolerate outsiders. Otherwise, they worked with the confidence that they could get to any team.

The success and international adventure cloaked a brewing darkness. Perumal brought the Zimbabwean national team to Malaysia for a pair of friendlies in July 2009. Zimbabwe lost the first match to Malaysia, 4–0. The syndicate profited from the fix, but the players were dissatisfied. Perumal hadn't paid them. During halftime of the second match, also against Malaysia, the Zimbabwean players argued with him. In the visitors' locker room at Shah Alam Stadium, one of the players said, "We're not coming out for the second half." Perumal pulled his coat aside. According to investigators, a handgun was tucked into the waistband of his pants. "You may not be coming back from Malaysia," he said. The players returned to the field for the second half, losing 1–0. Soon after the match, word spread that this was not the Zimbabwean national team, but a Zimbabwean club team called Monomotapa United. Perumal was taking new risks.

No location was too remote, no climate too extreme. Meganathan Subramaniam, who went by the theatrical alias of Simon Megadiamond, had, according to Perumal, been watching over the syndicate's operations in the Veikkausliiga, the Finnish first division. Perumal joined him at the grounds of the club Rovaniemen Palloseura (RoPS), located along the Arctic Circle, and there Perumal enacted a trusted strategy. Subramaniam had built a relationship with Christopher Musonda, the team captain, a Zambian who already had an acquaintance with fixing.*

On the afternoon of September 24, 2009, Musonda rode his bike to a fishing hole along a river outside Rovaniemi. Musonda stood out in this whitewashed town along the Arctic Circle. He had emailed a picture to his family. Standing in front of a towering snowbank, he held a chunk of ice in his hand. He was a long way from home. But he wasn't alone. Four other players for the RoPS club were also from Zambia, as was the team's coach, Zeddy Saileti.

The RoPS captain, Musonda, pedaled up to the riverbank with two of his teammates. Saileti was already there, accompanied by three men whom Musonda didn't recognize. When

---

* Meganathan claims that while he knows Perumal, he has never been involved in match fixing, that he is a simple taxi driver in Singapore.

the men spoke, in English, they did so with what sounded to Musonda like Eastern European accents. RoPS was scheduled to play a match against a club called TPS in a few days. Saileti explained that these three men wanted to make sure that the players understood what was supposed to happen. RoPS had to lose by a score of 4–2, or 3–1, or 3–0.

Saileti had driven to the fishing hole in a van. One of the Eastern Europeans, a large bald man, led Musonda inside the van, where he gave him €4,000, promising another €4,000 once the game was over. "But if the final score isn't what we expect," he told Musonda, "I'm going to sink all of you in this van in the river." The man wasn't laughing. Even though Musonda badly needed the money, he wanted to give it back. In the days leading up to the game, he couldn't sleep, "thinking what might happen if we did not lose the match." To Musonda's relief, RoPS lost comfortably, 4–0. When Musonda and his teammates received an additional €1,000, instead of €4,000, they didn't complain. They realized that they had been used, but there was nothing they could have done. Worse yet, they had gotten a taste for easy money.

On October 16, 2009, a day before the last match of the Finnish season, Perumal, Subramaniam, and Musonda met in a hotel room. Perumal wanted to lure Musonda away from the fixing group that he had met at the river. There was considerable action around the Finnish league because Finland was one of the few places in the world where soccer is played in the summertime. It was the only action in a quiet market. Perumal proposed to Musonda that they cooperate in fixing the entirety of the RoPS 2010 season. Perumal gave Musonda a "Christmas present" of $10,000, a gift designed "to support the cooperation in the coming year." It was a standard ploy, part of Perumal's charm. "You don't go directly at a player and try and set up a fix," he explains. "You take him out shopping, maybe. You buy him some clothes. You buy him some football boots. If he doesn't go for it, you write it off as a business expense."

CHAPTER 14

The 2010 World Cup presented Perumal with a unique opportunity. The betting market would be highly liquid for this month of matches. Even for the exhibition matches leading up to the main competition. On April 10, 2010, Football 4U International drafted a letter to Kirsten Nematandani, the president of the South African Football Association (SAFA). It was two months before SAFA would host the World Cup. The letter read:

> *Dear Sir,*
> *RE: REFEREES EXCHANGE PROGRAM*
>
> *With regards to our meeting on the above mentioned matter in Johanesburg [sic].*
> *Please note that we are extremely keen to work closely with your good office on the referees exchange program as discussed.*

We will be inviting South African FIFA qualified match officials for international friendly matches and league matches in Middle East countries whereby we act as agents for the supply of referees and assistant referees.

And in return we will be pleased to assist SAFA in providing FIFA qualified referees and assistant referees from CAF to officiate warm up international from May 2010 to June 2010.

We will also be glad to work with SAFA in organizing youth our tournaments and friendly international matches in the near future.

The letter was signed by "Wilson Perumal, Events and projects Executive." On April 29, 2010, according to a FIFA investigation, Jason Jo Lourdes arrived at the SAFA office at Johannesburg's Soccer City Stadium, the site of the eventual World Cup final match. Identifying himself as "Mohammad," Lourdes presented the April 10 letter to the SAFA receptionist, then asked to speak with the head of the referees department.

Despite the fact that it was about to host the most lucrative sporting event in the world, SAFA was in financial straits. SAFA had only just hired a new CEO, Leslie Sedibe, who was struggling to land the sponsorship deals that would fund the sort of training that the South African national team sorely needed. (Sedibe's fears were well-placed. South Africa was eliminated in the first round, the first World Cup host to fail to advance from the group stage.) The financial stress caused a rift in SAFA's hierarchy, which made the organization a perfect target for Perumal.

At Soccer City, Lourdes met Steve Goddard, the acting head of SAFA's referees department. After a brief conversation, in which Lourdes outlined the boilerplate Football 4U promotional offer, Goddard introduced him to Lindile "Ace" Kika, the head of the South African national team. Lourdes then spoke with Dennis Mumble, the chief competitions officer of the upcoming World Cup. Mumble also listened to Lourdes's pitch, then suggested that he meet with Sedibe. Within an hour after his arrival at Soccer City, Lourdes was having a closed-door meeting with the head of SAFA. Never mind that everyone in the soccer world had plenty of reason to be cautious about a Singaporean man appearing at the office, unannounced, with promises of riches. Never mind that the world would soon be watching SAFA's every move. To SAFA, an association with Football 4U appeared to be a sound idea.

At the fifty-seventh FIFA congress, held in May 2007, the organization ratified a new series of statutes. Among them was Article 13, paragraph 1(e), which states that each FIFA member association—that is, a national federation such as SAFA—must have a referee committee. The statute also delineates that this committee, under the guise of the greater national federation, shall be responsible for selecting referees and assistant referees for federation matches. The May 11, 2010, agreement signed by Leslie Sedibe and Wilson Perumal ceded SAFA's authority for appointing refs to Football 4U, in violation of FIFA statutes. But a statute is only as valid as the enforcement of it.

With this deal and others in place, Perumal was preparing for the World Cup. He and Dan Tan were back in business. But Dan Tan had some sobering news. He had four European partners. All expenses and all profits were shared equally among them, five ways. Even if Dan Tan wasn't involved in a fix perpetrated by one or more of his partners, he still earned. "This World Cup is divided by five," he said. Perumal believed he was being undercut. "If his partners do a game," he reasoned, "I don't get a share, but he gets a share. And if I do a game, they all get a share." Perumal went along with the plan. He had no choice. But he began plotting how to get compensation.

There was so much money circulating, and so little accounting of it, that Perumal recognized a simple way to dip into Dan Tan's pocket. Through the Hawala system, Dan Tan and the other heads of the Singapore syndicate would send Perumal the funding he needed to fix a match. Perumal simply padded the budget. "If the price was three hundred thousand dollars, I would quote Dan four hundred thousand," he says. As his gambling debts mounted, Perumal needed all the cash he could collect. The only way that he could get back in the black was to short the syndicate itself. Perumal would sometimes fail to fix a match, instead keeping the money for himself and hoping for the proper outcome. He had a ready excuse: many fixed matches didn't work out properly anyway. The ball was round. He might also run a fix with two clients, taking money from both of them, running one fix but not the other. This was a short-term strategy, however, as financiers would eventually stop throwing money at a gamble that produced no returns. Himself, Perumal couldn't stop gambling, nor losing. Before long, Perumal was in deep, $1 million in debt to one boss, $500,000 to another. Although tens of millions of dollars had trafficked through his hands, Perumal couldn't understand where it had all gone. But Dan Tan had a clue. Perumal had placed an underling in Egypt, in order to monitor operations in the domestic league there. When this man did the math, he understood that Perumal was skimming off the top. Judging Dan Tan a better bet than Perumal, the man explained to the boss why his wallet was lighter than it could have been.

CHAPTER 15

On July 17, 2010, six days after the close of the World Cup, Perumal was in Helsinki. It was the first week of the Finnish season, and Perumal summoned Musonda to his room at the Cumulus Hakaniemi hotel. RoPS was scheduled to play the Viikingit club the following day, and Perumal asked if Musonda could arrange for his club to lose. Musonda said that, on the contrary, RoPS could win easily, by a score of 2–0 or 3–0. The following day, before the match, Musonda and four of his Zambian teammates returned to the hotel, where Perumal gave them €25,000, half of their fee.

Musonda was a poor predictor of his team's performance. RoPS lost, 3–0. Perumal had wagered €200,000 on the match, and that money was gone. He met the players at the Helsinki train station, before their trip back to Rovaniemi. He angrily took back the €25,000 that he had fronted them.

He paid Musonda €60,000 for a 4–0 win against Tampereen Pallo-Veikot on July 25, 2010. RoPS won, but by a score of 3–0. Directly after the match, Perumal met Musonda in a men's room at the Tammela Stadion, where he took back the €60,000.

A week later, Perumal sat around Musonda's two-room Rovaniemi apartment with a handful of RoPS players, ultimately paying them €85,000 to forge a scoreless tie with Palloseura Kemi Kings, which they did. This match was one of Perumal's few successes with RoPS that season, during which he routinely gave Musonda a deposit, only to take it back after a botched fix. The money flowed in and then out of Musonda's hands, but he didn't mind. "If we had got the money for winning, it would have been easier to accept," Musonda would say later. "But now that the money would have been given for losing, losing the money felt like nothing."

Perumal never traveled alone, and was usually accompanied by an Arab and a Vietnamese man. The three would stand in the crowd at the 2,500-seat Rovaniemen Keskuskenttä stadium, wearing blue hats. Perumal would monitor the betting market on his smartphone, determining the best time to initiate the manipulations of the match. The three men would take their hats off, put them back on, and turn them to one side or another, communicating to the players a particular prearranged order. But sometimes the opponent simply couldn't score. When the fix did come off, it gave Musonda a funny feeling. "At times it was difficult to keep from smiling although we had lost," he said. It was a bruising experience, but Perumal had no plans to leave Finland. The Singapore syndicate was interested in buying RoPS, and the job of brokering the purchase had fallen to Perumal.

CHAPTER 16

Security may have been an afterthought for FIFA's executives, but they understood that it was important. Match-fixing? This was a marginalized issue, exotic, difficult to understand. Although he knew very little about fixing at the time, Eaton had a different opinion. He was a cop, and this was a crime. And as fixing was a crime of fraud, he understood that it had the potential to upend the very business that everyone else in the Zurich headquarters was so busy building and promoting. "I had been looking for a way to make a difference for some time," he says. "When I understood the scope of match-fixing, and the fact that no one appeared to be taking it seriously, I saw this as an issue that needed real leadership." At FIFA, Eaton was rejuvenated. As he sat there in his new, dank office, the episode with Boksic fresh in mind, Eaton realized that this was his last chance to achieve a goal of importance.

When he had left Interpol, Eaton had been concerned

about entering the private sector for the first time as he approached sixty years of age. Of greater concern, however, was how he would adapt to assuming a role secondary to his employer's mission. Whereas Interpol was a security organization, FIFA was a sports and promotional company, with security in a supporting position. With the match-fixing investigation blossoming into a meaningful project, Eaton, by force of personality and moral authority, created for himself a prime role. FIFA execs didn't know what to do about it. Eaton's job was security, not integrity. But Eaton judged that fixers weren't threatening to blow up stadiums, just the results. While FIFA looked to Eaton to oversee security for marquee events, like the upcoming announcement of which countries would host the 2018 and 2022 World Cups, all Eaton wanted to talk about was Boksic and Perumal.

Most everyone at FIFA had gained his or her position by couching the truth in palatable language, not by telling it like it is. Eaton's forthright tone was so out of keeping with FIFA's corporate manner that no one at FIFA dared challenge him. They also knew that he was right. When enterprising journalists in Europe and Asia and Africa began piecing together the picture of fixing on their own, they asked Eaton for comment, and he wasn't shy in providing it. Eaton had infiltrated FIFA, in a way, and he was now forcing the organization to reassess itself, as he posited a new argument: if the individual games were corrupt, then the entire game was called into question. Fixing was bad for business, but was Eaton worse? Whatever the answer, both publicly and within the FIFA bunker, it had become impossible to ignore Eaton.

FIFA execs should have understood that this was a likely outcome. Prior to his employment, Eaton was compelled to sit for a lengthy psychological review, as was customary for prospective employees. This was a two-day test administered by two German doctors. Their assessment of Eaton's personality confirmed what everyone at Interpol already knew. In part, it read:

"Mr. Eaton has a noticeable tendency to be too full of energy in doing his part. This creates the danger that he overwhelms others with his dynamism. . . . One has to be able to put up resistance against him; otherwise one runs the risk of being relegated to the background. . . . His decidedly quick cognitive ability allows him to see the big picture and understand what is important." The report may have been useful. But it was more to the point to describe Eaton as an Australian colleague later did: "Chris is like a general in the Boer War." Eaton was the sort of senior-level independent actor that any company would both covet and fear.

CHAPTER 17

Through Football 4U, Perumal arranged a match between Bahrain and the tiny West African country of Togo. This was a classic arrangement of the Singapore syndicate, an impoverished soccer federation playing yet another meaningless international friendly on the edge of anyone's concern.

Perumal's favorite referee, Ibrahim Chaibou, was officiating. Access to the betting services was assured. There was no need to take chances. But from his earliest days, Perumal had been a gambler. His brazenness had been the key to his success as a match fixer, the ability to do what others found impossible or objectionable.

On the evening of September 7, 2010, the Togolese national team took the field at the national stadium in Riffa, Bahrain.

But this was not the Togolese national team. It was a team of journeymen club players from Togo, selected by a former national coach whom Perumal knew. The real Team Togo had lost to Botswana in an African Cup of Nations qualifier only days before. Togo lost to Bahrain, 3–0.

After the match, Josef Hickersberger, the Bahraini head coach, shook his head, commenting that the Togolese players "were not fit enough to play ninety minutes." This episode recalled Perumal's use of the questionable Zimbabwean national team the year before. The difference this time was the considerable media attention that the match generated. By now, the sporting press had caught on to the fixing story. Perumal had made Togo look bad. He had made FIFA look ridiculous, which, it turned out, required a special effort before an uninformed, forgiving public. The publicity put the syndicate at risk.

The only reaction Perumal had was to retire Football 4U, changing the company name to Footy Media. "The bubble burst after the Bahrain game," Perumal says. "Before then, I was very quiet, below the radar. After the Bahrain game, I became very popular." Otherwise, he didn't care. He believed so strongly in his invincibility that he had no qualms with living in the public sphere. He maintained a Facebook page. He posted photos of himself from various locales around the world, dated proof of his whereabouts, which any investigator would find useful. Perumal also had an account on Zorpia, a social network popular in China and India. His username was zidane3107, no doubt a reference to the French national team star. He listed his favorite book as *The Day of the Jackal,* Frederick Forsyth's 1971 cloak-

and-dagger novel about an assassination attempt on the French president. His favorite movie was *The Shawshank Redemption*. Under "Interests," Perumal noted just one: soccer.

Perumal didn't know when to stop, and what would have inspired him to do so? FIFA, the soccer federations, and law enforcement had never cared much about his behavior. The cops weren't closing in. But the syndicate was. Dan Tan had come to the conclusion that Perumal saw himself as the main character of a soccer thriller, whereas the other members of the syndicate were concerned only with running a business. The Singapore bosses recalled Prakesh and Anthony Raj Santia. At a meeting in Singapore, Dan Tan told them that he no longer trusted Perumal. He wasn't a sure bet anymore. His ego was out of control. The bosses wanted to continue the operation with people they could trust.

In a business based on betting, Perumal's gambling had impinged his credibility. And when he turned on the TV on October 8, 2010, he knew that his situation was about to change. Bolivia was playing Venezuela in an international friendly. The match featured two classic Perumal national team targets, but that wasn't what caught his attention. The thing that struck him was the identity of the man officiating the match. It was Ibrahim Chaibou. Chaibou had awarded three controversial hand-ball penalty kicks in South Africa's 5–0 win over Guatemala in May 2010. As Perumal tells it, he had annulled goals on phantom offside calls, extended injury time long enough to pad the score, issued red cards at critical moments in games, and it appeared, when Anthony Raj Santia contacted him about refereeing the match at Venezuela's Estadio Polideportivo de Pueblo

Nuevo–San Cristóbal, he went right along with the plan. There was no loyalty between fixer and the fixed. Pal Kurusamy had taught Perumal that long ago. But what about loyalty between fixers, between business partners? Santia and Prakesh had gone around Perumal, tapping into the network that he had spent years establishing. They bypassed their mentor and plugged into the Singapore syndicate directly. Perumal was an entrepreneur, sure. He could sell his fixes to whoever was buying. But he was also a shareholder in the joint Asian-European fixing cooperative that appeared to be shutting him out. He wasn't the sort of person to allow that to happen.

CHAPTER 18

Carsten Koerl grew up in Allgäu, a province of Bavaria, half-way between Munich and the Austrian border. His fascination with both sports and computer programming led him, in the mid-1990s, to investigate the ways in which he might combine his interests.

In 1998, with a friend, he founded a company called Betand-Win. It was one of the first bookmakers to offer betting through an online portal, a novel concept at the time. Koerl could not have been more prescient. Just four years after Koerl opened his company with five employees, several thousand people were working at BetandWin. In 2002, he took the company public. Its name shortened to Bwin, the company eventually became one of the dominant books in the market. Bwin went on to sponsor Real Madrid, AC Milan, and Bayern Munich, the company's logo emblazoned across the clubs' jerseys.

When Koerl removed himself from the daily operations of

the company, shortly after the IPO, he cast about for a new project. He remembered two Norwegian programmers who had pitched him on a product during his last days at Bwin. He didn't have time for them then, but now the memory returned to him.

Ivar Arnesen and Petter Fornaess were in their mid-twenties, graduates of the Norwegian University of Science and Technology, located in the city of Trondheim, three hundred miles north of Oslo. The university in Trondheim held mythic status among computer programmers. Engineers there had been instrumental in building the operational core of Google.

Like Koerl, Arnesen and Fornaess were interested in sports and computer programming. And gambling. They knew enough about the sports betting market to recognize its inconsistencies. Bookmakers were good at analyzing a match and offering the kinds of odds that would generate action and produce profit for the company. Setting the right odds was the bookie's special talent. What bookies weren't good at was paying attention to small details. And these small details could make a big difference.

Arnesen and Fornaess realized that bookies, since they were dealing with so many games, often listed the incorrect starting time of a match. Sometimes match administrators moved a contest from one site to another, a change that could affect the outcome, yet a bookie might fail to report this fact. A bookie could receive the wrong information from a spotter and list the incorrect final score of a game. Also, without noticing it, a bookmaker might list odds that were, say, 15 percent higher or lower than the market average. If a bettor possessed more accurate information about a match than did the bookie with whom he was wagering, then he had the upper hand.

Bookies were taking action on league matches, international friendlies, World Cup qualifiers, youth games—from all over the world and at a constant pace. With so much volume, bookies simply couldn't keep track of every detail of every match that they listed. Arnesen and Fornaess developed a method by which they could. They created an algorithm that could handle mass data extraction, specific to the online betting market. Using the data supplied by the online books themselves, their program was able to pinpoint inconsistencies in the market. Arnesen and Fornaess could identify the weak link (or links), the outlier in the network of international bookmakers, and then exploit it. When they detected which book was listing the wrong odds on a particular match, they would place a bet on that game with that book. They had figured out a way to game the system, using one book against many others to make a profit.

As exhilarating and profitable as this was, Arnesen and Fornaess understood that as the market became more developed, which was an inevitability, bookmakers would gradually address and eliminate these inefficiencies. Eventually, Arnesen and Fornaess would use their algorithm to squeeze out smaller and smaller profits, until their tool became virtually worthless. They believed that there was a better, more sustainable way to utilize their program. They decided to develop an instrument to aid bookmakers in correcting their offerings.

They pitched the idea to Bwin. But they weren't businessmen. They were programmers. Their product and pitch were raw. All the same, Koerl recognized an opportunity, for he was the entrepreneur. And after Bwin's perfectly timed IPO, he had the abundant capital to finance a start-up.

In 2004, Koerl, Arnesen, and Fornaess established a new company, calling it Sportradar. The company's product, which was called Betradar, expanded upon the technology that Arnesen and Fornaess had been using to beat the bookies, but with Koerl's new market-specific modifications. "From my time at Bwin, I knew what a bookmaker needed, because I was a bookmaker," Koerl says. "I understood risk management, because I learned about it in a painful way, from clients cheating on me. I had precise knowledge of what I needed. That was our advantage. This was something that the bookmaker needed." For a monthly fee, which a bookie could build into the back end of his business, like insurance, Betradar would handle the tedious tasks of verification and data collection for which bookies traditionally displayed little talent or patience. Koerl was back in the bookmaking game, yet in a decidedly different role. It wasn't long before that role would change once more.

In January 2005, German soccer officials announced the suspension of referee Robert Hoyzer. During an August 2004 German Cup match between Paderborn and Hamburger SV, Hoyzer had awarded underdog Paderborn two questionable penalty kicks. He had also dispatched a Hamburger player by red card. Paderborn won by two goals. An investigation determined that Hoyzer had bet on the match.

As the German soccer federation combed through Hoyzer's match record, the German prosecutor's office initiated its own investigation. It turned out that Croatian organized crime had been controlling Hoyzer, who had fixed a number of matches. And he wasn't the only one. Revelations implicated dozens of players and referees, even hinting that someone from inside UEFA was leaking the identities of those who would referee Champions League matches, allowing the Croatians to compromise them. Hoyzer was sentenced to two years and five months in prison. The leader of the betting ring, Ante Sapina, received two years and eleven months. The case was closed, but Koerl knew that this was not an isolated incident.

From years working in the betting business, Koerl knew that fixing was happening with regularity. Every bookie knew. They could usually tell when something untoward was happening with a match. It was one of the challenges of the business. Not only was a bookmaker concerned with establishing the proper odds to offer his clients, but he also had to worry about fraud in the marketplace upending all of his calculations and ultimately cleaning him out. Everyone in the business had general knowledge of fixing. What made Koerl different was his specific knowledge. He just didn't know that he had it.

Shortly after the Hoyzer case broke, Koerl received a phone call from a journalist. The reporter informed him that several bookies had revealed that they had recognized these fixed German matches through the Betradar interface. One component of the software was called "removed lines." This displayed bookmakers that had offered betting on a match, but then subsequently removed it. This had always been a curious practice, for if a bookie removed a match after first listing it, he must know that the match was tainted. Combining this element with other factors of the Sportradar software, Koerl approached the German Football Association.

He explained that his technology could alert the federation to the likelihood of a fixed match. And it could do it in advance of a game, so that the federation could take action to protect its organizational integrity. When the German Football Association, mired in scandal, enthusiastically signed a contract, Koerl and his team set about modifying the technology. The new product had to be designed not for bookies, but for soccer administrators.

Koerl's new system analyzed the numerous characteristics of a given match—its location, the relative strengths of the opponents, player injuries, the weather, and many more data points—in order to produce the most logical betting line. This was the work of a bookie, establishing a baseline. Then Koerl's system compared this ideal, baseline bet to the actual offers of the sportsbooks. When the market offers deviated wildly from the established baseline, this was a signal that a match was questionable. Koerl knew that there were only two reasons why a bookie would move his numbers drastically from what they should be. Either the bookie had inside information, or he had taken big action that he didn't want—that is, bets on a fixed match. For example, if a team's chance of winning is initially 45 percent, and then a bookie pegs the chances at 70 percent, the game is undoubtedly fixed.

The program served a function similar to the analysis of insider trading on a stock market. If the price of a share suddenly collapses for no apparent reason—the CEO hasn't been fired, the quarterly earnings report contains no surprises—there is likely some nefarious explanation for the drop in value. Koerl's system was able to review the share price of a soccer game, then investigate why it had collapsed. He recognized that he could see the marketplace deteriorate ahead of time.

Koerl named his new product the Early Warning System (EWS). Beginning in 2005, with its first client, the German Football Association, EWS monitored five levels of German soccer, from the Bundesliga on down, roughly five thousand matches a year. Koerl then forged deals with the football federations of the Czech Republic, Estonia, and France. When UEFA signed on as a client shortly thereafter, Koerl's new venture attained unquestioned credibility. EWS was now overlooking every professional soccer match in Europe, from fifty-four member states. This arrangement enabled EWS to analyze not just the activity in one country, but the ways in which fixing in one country influenced fixing in another, and how betting was being placed in Asia. By the time FIFA signed on, in 2007, EWS was the recognized authority in its field, with Koerl supporting governments and police across Europe. It had taken a few years, but soccer administrators had come to realize that the EWS was filling a necessary gap in their security.

"Manipulation is damaging to the commercial possibilities of sport," Koerl says. "Federations are very aware of the connection. It's a massive commercial industry." Altruism aside, protecting the integrity of the games with EWS supported Sportradar's core product, Betradar. "Do you think sports betting would exist if people knew the outcomes of the game before it starts?" Koerl asks. "I think not." There was just one complication. Koerl's clients didn't know it, but by 2007, his EWS was obsolete.

Only insiders understood that by 2007, technological innovation had fundamentally transformed the betting market. Criminal syndicates were now able to manipulate bookmakers with impunity, hitting this corner candy store over and over. And there was no way to detect their crime. Koerl understood that he would have to adapt his technology to the new reality. When FIFA developed its own in-house monitoring algorithm and called it the Early Warning System, Koerl let them have the name. In sports betting, there was no such thing as an early warning anymore. "The name was not really appropriate for the kind of manipulation that had developed based on how popular live betting became," Koerl says.

Live betting, or in-game betting, existed before 2005. But 2005 was the year that it began to affect the marketplace in a significant way. Traditionally, a bettor placed a wager on a game before it started. And he was stuck with the bet. If the game's events began to affect his chances adversely, there was nothing he could do. He couldn't relinquish his bet. He couldn't make a new one, since the game was already under way. He might yell at the TV screen, but that was the extent of his ability to express himself.

By 2005, technology was beginning to enable the bettor to monetize the thoughts he was having about the flow of play in a live match. Broadband Internet had penetrated deeply into homes in Europe and Asia. Smartphones were taking over outside the home. Instead of having to walk down to the bookmaker's shop to place a bet, a bettor could bet on his computer at home, or on his phone while at the bar watching a game. Online portals like the one that Koerl had pioneered at Bwin began taking abundant action.

And the kind of action they were offering changed in accordance with the new form of betting. Bookmakers realized that online betting allowed them to provide a new array of offers, stimulating the betting market and driving liquidity.

The types of bets didn't change. You could still bet the Asian handicap, the over-under, and other wagers familiar to the traditional client. What in-game betting changed was how frequently the dynamic of these bets changed, as a match progressed. Any event in a match—a goal, a red card, or simply expiring time—affected the game itself, and thus altered the probability of the bets related to its outcome. As a match played out, a bookmaker would now change the odds he offered on a bet depending on the shifting factors of the game action. If a bettor realized that his chances of winning a wager that he had already placed were decreasing, he could now bet against himself, or he could place another bet entirely. He was no longer stuck with the bet he had made before a game. And he could place bets throughout a match. Suddenly, sports betting changed from a static, passive experience—a fan's experience of watching the game—into a frenetic opportunity for arbitrage, stimulating the adrenaline that one might feel by playing the game. Bettors converted to the new way in droves, fundamentally altering the betting landscape. In 2003, 10 percent of Bwin's online sportsbook business was in-game betting. By 2010, in-game betting represented 70 percent of Bwin's business.

Sophisticated criminals are quick to take advantage of innovation. The match-fixing syndicates found enormous potential in the advent of in-game betting. Previously, they had to guard against placing pregame bets so high that they would alert bookmakers to an upcoming fix. Now they could wait until a match began, then place their bets during the action. Once a game had started, a league or federation wasn't going to stop it. And if the syndicates placed their bets judiciously, it would be too late before the bookie could detect them, if ever.

Live betting also placed a fixer in the stands at a game, smartphone in hand, monitoring the shifting odds of numerous bookies, communicating with his players and referee. These constant line and odds changes stimulated betting, invigorating the market. There was no innovation that could have better enhanced the business of fixing.

Carsten Koerl alerted his clients to this new reality, to the dangers of live betting. He also told them that he had developed the tool to combat it.

Sportradar currently employs 650 people worldwide. Two hundred IT experts code, develop, and maintain the infrastructure. Another couple of hundred operatives manually collect data from matches. There are roughly 150 journalists on stringer agreements to provide the minutiae from remote locales and clubs. To the games themselves, Sportradar dispatches roughly two thousand scouts, who for a nominal fee transfer match info directly by touch-tone phone keypad. All of this data goes into the system.

In 2009, Koerl launched Sportradar's new product, the Fraud Detection System (FDS), which analyzes the shifting data from a game in order to identify a fixed match, real-time. The system originates its baseline odds based on three parameters: time elapsed and remaining in a game, the number of goals for each team and total, and the number of red cards. This allows the system to calculate the true probability of a team's winning a match. The FDS then overlays this against the offers of 350 bookmakers, measuring the differences. When the deviation between the baseline offer and the actual offers grows too great, the system issues an alert to Sportradar's analysts. If a match is considered suspect, analysts assign a warning level to it, yellow for suspicion, red for certainty. Everything happens instantaneously.

"Bookmaking is not a genius sitting in a dark chamber coming up with odds," Koerl says. "That doesn't happen anymore. It's driven by algorithms, mathematicians, massive progression models. It's much more related to financial markets."

And as the FDS has progressed away from the individual, it has become more personal. Prosecutors across Europe frequently rely on Sportradar's data in fixing cases. Through these experiences, Koerl has realized that what prosecutors and judges value is not raw data or endless analysis, but evidence that implicates players, refs, and coaches. "We learned that the system needs to deliver more information about individuals," he says.

Sportradar developers added features that enabled the FDS to identify and track suspicious actors. If one club had participated in manipulated matches, and one or more of its players ultimately moved to another team, the FDS would make a notation that allowed a system user to collate related information. The FDS goes into great detail, pinpointing even a player's performance in a game, a season, or a career. If a defender causes two penalty kicks in a match, for example, the likelihood of his being involved in a fix increases, a fact duly noted in the system. With 190,000 personal profiles in its database, the FDS is the Big Brother of match-fixing.

But what good is all of this data? Some soccer industry professionals complain that Koerl and his people at Sportradar are a mop-up crew. Once the fixing syndicate strikes, it is gone, it has won its bets, and the game is over. The FDS is not a system of prevention. It is retrospective, even as it clocks real-time progressions in the market. Some federation officials complain about Sportradar's voluminous reports about fixed matches, the paperwork filled with data and figures that mean little to anyone but an IT professional.

Conversely, Koerl claims that he has provided some of his clients with information that points unquestionably to the fraudulent activity of players in their leagues, to no avail. "Sometimes I'm not happy with some federations, how they integrate our information, how they go about using it," he says. "I tell them the right thing to do, without political intentions." The recognized authority in his field, Koerl has been searching for the landmark criminal case that would prove Sportradar's prescriptive value.

CHAPTER 19

ANTALYA, TURKEY, FEBRUARY 2011

Sportradar was scrutinizing events in Turkey. The Mediterranean resort town of Antalya was hosting two friendlies: Latvia versus Bolivia, and Estonia versus Bulgaria. This was another round in the endless series of exhibitions that held relatively little significance in any standings. This inconsequence made matches of this nature particularly vulnerable to manipulation, though that didn't stop bookmakers from offering odds on them. Invariably, any game between national teams drew mass liquidity in the global market. Recognizing the fact that large numbers of bettors were eager to wager on a game for which the players themselves didn't care, was elemental to understanding how the fixing syndicate was able to strangle the sport. That was the criminal genius behind Perumal's fake Togo match. He realized that it didn't matter who wore the national team jersey. People were going to bet regardless.

Footy Media International had arranged the Antalya games (in the name of Blapp Johnsen), and in a way that immediately aroused the suspicions of Janis Mezeckis, the general secretary of the Latvian Football Federation. On behalf of Footy Media, Anthony Raj Santia had traveled to Riga and offered the federation €30,000. He said that his company would arrange travel and accommodation for the Latvians, as well as the rental of the stadium and the selection of match referees. Mezeckis signed a contract with Footy Media on December 14, 2010. But he and his colleagues questioned how Footy Media planned to profit from the matches, as TV and sponsorship rights were not included in the proposal. Santia claimed that his company was eager to build its profile in Europe and so viewed this deal as an investment in future business.

It didn't add up for Mezeckis. He contacted FIFA. Chris Eaton alerted Sportradar. The Antalya matches went ahead as planned, one of the most egregious episodes of match-fixing ever conducted.

As game time approached on February 9, 2011, Janis asked Santia to confirm the identity of the match officials. Would they be from FIFA's approved list? Santia had assured him that they would be, and that the officials would be Czech. One hour before kickoff, Santia informed Mezeckis that the refs were Hungarian. Mezeckis walked down to the referees' dressing room in Antalya's Mardan Stadium. There, the officials said that they were, in fact, Croatian. In actuality, they were Bosnian.

Darren Small, the chief operating officer of Sportradar, watched the betting for each game in Antalya behave identically. The pregame over-under was 2.5 goals, meaning that if you bet the over, and the two teams in a match combined to score three or more goals, you would win. This line generated massive betting on international bookmaker sites. Once the first goal of each game was scored, most bookmakers that Sportradar monitored pushed their over-under offer to 3.5, and the betting action subsided. With scoreless time passing in each game, the line dropped on most books, returning to 2.5. The betting action increased once more. "There was severe support for more than three goals in both games," Small says. "The support in the marketplace for those games was nowhere near logical, with about five million euros bet per match on the over-under."

But it wasn't the betting movement that singled out these matches. It was how the results were achieved. There were at least three goals scored in aggregate in each game, as the betting patterns suggested there would be. Together, the games generated seven goals. This would be unremarkable but for the fact that all seven were scored from penalty kicks. In the case of one penalty kick, after the player missed, an assistant referee ordered that it be retaken.

Although Chris Eaton and his subordinates believed that Perumal was behind the Antalya matches, this was not the case. Perumal was watching from afar. When he learned about the way that the fix had transpired, he laughed. "How dumb can you be to play seven goals in a match?" he says. "You're better off if you flash one red card to that team, and make sure the other wins by two–nil. So it doesn't seem so awkward. And it's not so obvious. Seven penalties. Same venue. Same day. Same company. Something is very wrong. Dan Tan destroyed a lot of things."

Perumal believed, rightly so, that an obvious fix spoiled the business for everyone in it. He had been guilty of the trespass himself, with the fake Togo match. His resentment was focused on Dan Tan, who he believed was better in the background, rather than in operations. The two of them had arranged a December 2010 fix in Córdoba, Argentina, between the under-twenty national teams of Argentina and Bolivia. The plan was for Argentina to win in the last five minutes of the game. The referee disallowed a legitimate goal in the thirty-fifth minute. Regulation time expired with no further scoring. When the ref awarded Argentina with a penalty (successful) in the twelfth minute of injury time, it made a mockery of the game. Perumal claims that Dan Tan got greedy. "You have to be prepared to lose at some stage, and not win every fucking spin of the roulette," he says. "You are fixing. One time it can fall on the wrong number. When they scored this twelfth-minute injury time penalty, the whole of South America came to know about match-fixing. Dan Tan completely destroyed our business in South America. He had this feeling he was bulletproof, that nobody is watching him, that nothing is going to happen, that you have license to do this."

The obvious fraud in Antalya also insulted Chris Eaton's sense of propriety. Further, he believed that he wasn't dealing with sophisticated criminals, since Santia had done little to conceal the fix. Any child who had ever kicked around a ball in the backyard could tell that there was something wrong with these games. Yet Santia paid the refs. The players jogged off the field. And somewhere in the Far East, the syndicate was counting its money.

If a fix could be conducted so obviously and openly, Eaton realized, no amount of policing could defeat it. Or it would have already done so. A new approach was needed. Eaton began to think of fundamental reform, an institutional shift, a new set of regulations that would infiltrate the syndicate, protect players, and steadily dislodge the fixers from their entrenched position.

And Eaton thought of Perumal, and how Antalya had come and gone without him. "We thought we had Perumal there," he says. The chase would continue, though Eaton feared that it might go on forever.

CHAPTER 20

Perumal was angry over the fact that Dan Tan had left him out in the cold, but he kept his sentiments to himself. He needed money to cover his mounting debts. In Finland, he negotiated on behalf of a Hungarian front group, agreeing on a $500,000 sale price with the Finnish owners of the club. When Perumal received this cash amount from syndicate runners, he gave the Finns only $200,000, retaining the balance for himself. The Finns phoned the Hungarians. The Hungarians called Singapore. When everyone gathered in an office in Rovaniemi, it was clear who had been the cause of the misunderstanding.

Perumal's charm had begun to wear off in anger and irritability. By early 2011, relations between Perumal and his main corrupted player at RoPS, Christopher Musonda, were strained. On February 16, RoPS lost to the club Vaasan Palloseura by a score of 3–0, though Perumal had arranged with Musonda to lose 4–0. This enterprise was bleeding money. When Musonda returned to the locker room after the game, he received a text message from Perumal. "U guys are stupid," it read. "Where is the one more goal. So close n still cant get the job done."

It was a wonder why Perumal stuck with his Finnish club for so long. But by the end of February 2011, with the arctic winter having closed in, he had reached his limit. On February 23, RoPs drew a 1–1 tie against Tampere United, though Musonda and his teammates were supposed to have lost. Perumal met Musonda and two of his teammates at Fransmanni, a formal French restaurant near the stadium. Perumal scolded them. He was so consumed in the frustration that his business caused him that he failed to notice that he had been followed to Fransmanni. Perumal well understood that injuring the syndicate's business carried a price. The Singapore syndicate had finally made its determination about Wilson Perumal. It wouldn't be long before he knew what it was.

CHAPTER 21

The FIFA executive committee tabulated its results on December 2, 2010. When Sepp Blatter climbed the stage at FIFA headquarters in Zurich, he said simply: "We go to new lands." FIFA awarded the 2018 World Cup to Russia. The 2022 World Cup went to Qatar. Six months earlier, FIFA had staged its first African World Cup. After a stopover in Brazil in 2014, the fifth time that the tournament would be held in South America, FIFA would take the World Cup competition to Eastern Europe and the Middle East, the first time for both regions. This was a distinct acknowledgment of FIFA's regard for emerging markets. Or was it? As soon as Blatter had made the announcement, charges of corruption and vote-rigging began to circulate. And no wonder. A month before the World Cup ballot, FIFA's ethics committee had suspended two members of its executive committee—a Nigerian, Amos Adamu, and Reynald Temarii of Tahiti—suspected of attempting to sell their votes.

Qatar's poor reputation meant that the preponderance of the accusations was directed at it, and not at Russia. FIFA's own scouting report on Qatar's viability as a World Cup host warned of the dangers of the high summer temperatures, which were a health risk to fans and players alike. Members of the U.S. contingent, which lost to Qatar in the final round, had difficulty understanding how the United States' developed infrastructure, large population and fan base, and the marketing opportunities of its economy failed to attract enough votes. Those familiar with the history of FIFA's internal politics recalled the organization's contentious 1998 presidential election. At the time, there were whispers that the Qatari government had financed Blatter's campaign, enabling him to defeat the reformer Lennart Johansson, a Swede, who had once appeared to be the favorite. Insiders now alleged that Mohamed bin Hammam, the Qatari member of FIFA's executive committee, had paid for votes. But was there any proof? The World Cup brings great financial benefit to the host economy and those closely tied to it. The losers in the World Cup selection process routinely cry foul. FIFA's murky histories make it an easy target.

Only the rare FIFA vote—whether it was to determine a tournament host or who would be president of the organization—was devoid of suspicion. From João Havelange's 1974 ouster of Stanley Rous as FIFA president, to the voluminous, opaque licensing and sponsorship deals that Blatter had engineered, it was often hard to tell how things transpired. FIFA didn't benefit from the sorts of financial controls that usually guide large organizations. The photo of Blatter with the infamous Russian organized criminal leader Taiwanchik (who was charged with fixing the figure-skating competition during the 2002 Olympics in Salt Lake City) may have been an unfortunate coincidence for the FIFA president. But it served to confirm ever-growing public sentiment that Blatter and FIFA were unscrupulous. It felt as though there was always another FIFA scandal waiting to happen.

That's the one benefit often overlooked in a state of continuous scandal: a new disgrace obscures the one previous to it. No sooner was the World Cup announcement completed than another FIFA election materialized in the near distance, to be held on June 1, 2011. Blatter's four-year term was about to expire. Surprisingly, in the upcoming election that would determine if he would stay or go, Blatter was facing a serious challenger. The identity of this candidate came as a shock. Mohamed bin Hammam had always been a close ally.

A longtime member of FIFA's executive committee, bin Hammam also sat on Qatar's Advisory Council, a select group that counseled Sheikh Hamad bin Khalifa al-Thani, then the country's emir. Bin Hammam was close to power, and he wanted some of his own. On March 18, 2011, he announced his candidacy in the upcoming election. In the wake of the World Cup announcement, Qatar was gaining momentum.

CHAPTER 22

Eaton dispatched an alert to various national soccer federations, advising them to be aware of bogus front companies, of Tamils from Singapore, of a man who called himself Raj. While concerned about the vulnerability of these organizations, Eaton privately hoped that Perumal would attempt to compromise one of them, thereby revealing himself and his location. Eaton continued to compile evidence from the many global sources he had accumulated during his Interpol days. To him, these pieces composed a conspiracy, with Wilson Perumal at its center. But Perumal was the one piece that was missing.

Eaton felt like he was the only one in Zurich who understood the importance of finding Perumal. This was another sign of Eaton's incompatibility with his colleagues. He took no part in the "team-building" activities outside the office. He knew that he would never fit in with the reserved, Swiss way of doing things. When he did make jokes with people around the office, as was his way, they often took him at his word. Eaton concentrated on his operational task, rather than on internal politics. Invariably, he took his lunches alone. On March 1, 2011, having recently returned from San Salvador, he returned to his drab basement office after another solitary lunch. His phone rang.

On the line was Graham Peaker, intelligence coordinator at UEFA. He and Eaton had cooperated on previous match-fixing-related inquiries, sharing information and expertise. Peaker specialized in IT. He wasn't an investigator. But he knew that Eaton was. He was calling now because the phone request he had just received required Eaton's experienced hand. Peaker said that he had information about a crime that had been committed in Finland.

Eaton was interested. He listened intently to Peaker. "Police in Finland have arrested someone," Peaker told him. "We don't know who he is. But he has been fixing matches up there."

Eaton asked Peaker for details. Until now, Eaton had been dealing only with compromised players and soccer officials. The fixers themselves were phantoms, aliases signed at the bottom of an email, blurred images on stadium security videos, the binary code of mounting data. Peaker said that he would supply more information in an email. But Eaton persisted. He had a feeling about this one, and he wanted to confirm it right away.

"Okay," said Peaker. "The police said that they observed

this guy yelling at the players and threatening them." Eaton sat up in his chair, excited. Could this have just fallen into his lap?

Peaker said: "The players call him Raj."

Eaton knew that millions of Indians went by the name of Raj. But there could be no mistake. This Raj had to be the one he was looking for. Eaton told Peaker: "I'm going to Finland."

CHAPTER 23

Nine days earlier, on February 20, 2011, a man named Joseph Tan Xin walked into the central police station of Rovaniemi, Finland. Tan Xin, a Singaporean national of Chinese descent, informed the duty officers that another Singaporean man, of Indian background, was currently in town traveling on a false passport. Tan Xin identified the man as Wilson Perumal. When the officers asked Tan Xin why he had supplied them with this information, he reacted defensively, saying that he would leave the station if they treated him as a criminal.

Jukka Lakkala, an inspector with the Rovaniemi police, walked over to the Scandic Hotel, where Perumal was lodging. There Lakkala and a colleague took Perumal into custody. When they returned to the station, the officers realized that they had detained the wrong suspect, another Indian man who had been staying in a different room in the same hotel. Careful not to make the same mistake twice, Lakkala placed the hotel under surveillance. When Perumal appeared, police followed him.

On February 23, they watched Perumal join three other men, European in appearance. The group headed to the Keskuskenttä stadium, where RoPS was about to play Tampere United. In the stands, Perumal held his phone to his ear throughout almost the entire match, talking only occasionally. Lakkala and his colleagues couldn't figure out what might cause him to behave this way. They were intrigued. They didn't know that Perumal had paid Musonda €80,000, 10,000 per participating player, for RoPS to lose by a score of 3–0. On the phone, Perumal was directing his betting.

A bookie in Macau had been hounding Perumal over a $200,000 debt. Via email, the bookie wrote that he knew Perumal was traveling on a false passport, that it would be easy to alert the police. Perumal was offended. That was against the criminal code. You never went to the police. And he was anxious. Any contact with the police in Finland could lead him to a prison cell in Singapore, where his five-year sentence awaited. But if Musonda and his RoPS teammates could affect the fix, Perumal would go a long way to solving his financial difficulties. In this small stadium, in this forgotten league, how hard could it be to influence the game's outcome? No one was watching. He had most of the players in his pocket. But of the seven matches that Perumal had attempted to fix with RoPS that season, only one had come through. Instead of stockpiling the money he needed to settle his debts, he was losing. This was the final match of the RoPS season, Perumal's last chance. On his phone during the match, he stood to win $1 million, but RoPS had to lose. When the game ended, Perumal couldn't believe it. RoPS had tied, 1–1.

Police were watching. They followed Perumal and one of his associates to Fransmanni, the French restaurant near the soccer stadium. Officers watched the restaurant as Musonda and two other players joined Perumal. It was a contentious meeting. The players cowered as Perumal and his associate appeared to threaten them. To Lakkala, this all looked more interesting than a simple case of a fraudulent passport. But he wasn't sure what he was watching. Although various fixers had been manipulating RoPS matches for at least the past three seasons, doing so through the club's most popular player, who had then become its coach, the local police professed to have no idea of such activity.

Two days later, at 6 A.M., Lakkala watched Perumal board a flight from Rovaniemi to Helsinki. Lakkala phoned ahead to his colleagues in the capital. When Perumal arrived at Helsinki's Vantaa airport, customs officers took him aside. They scrutinized his passport. Chelliah Raja Morgan was the name on the document. The date of birth was listed as October 7, 1987. Perumal may have been a youthful forty-five years old, but he could hardly pass for twenty-three. On February 24, 2011, at 8:50 A.M., police took him into custody. But they didn't know who he was.

Perumal was desperate to keep it that way. However, the trail of evidence he had left behind threatened to undo him. Once police transported him back to Rovaniemi, Lakkala and his colleagues began going through Perumal's belongings. They recovered files from his laptop. They read his emails and text messages. They looked through the numbers in his phone. His phone book included a Cyprus number for "Dan." Two of Ibrahim Chaibou's numbers were there also. There was a listing for "Dead Ball Specialist," as well as numbers for Christopher Musonda, Togo soccer officials, bookies, and criminal figures and financiers. In all, he had contacts from thirty-four different countries, from Venezuela to Slovakia to Qatar to Zambia. From Perumal's Rovaniemi hotel room, police recovered the scratch-pad math of his fixes, what he paid players, the amounts he expected to make on the betting market. The notes referred to a series of under-twenty-three friendlies that he was arranging in Kuwait, between the Kuwaiti national team and Syria, Palestine, and even Switzerland, Singapore's European equivalent. All told, the evidence constituted the most complete record of the Singapore syndicate that authorities had ever seized. But to the Finns, it was a thousand scattered puzzle pieces. They couldn't see the image they were looking at.

Although Perumal would boast that "a five-year jail term in Singapore is nothing," avoiding this stretch remained a guiding concern. As police led him, handcuffed, from a squad car toward the steps of the Rovaniemi courthouse, he broke free and ran. It was February along the Arctic Circle, twenty degrees below zero. Perumal was wearing a T-shirt and jeans. He was handcuffed. When he tired, cold and aimless, he stopped running. A cop cruiser pulled up alongside him. He got in.

If he was going to avoid that long flight back to Singapore, Perumal was going to have to make friends in Finland.

When Chris Eaton's Finnair flight landed in Rovaniemi, there was darkness all around. It was dead winter, twenty-four hours of night in a day. Eaton was in the right place. He had traveled to Lapland in order to enlighten Finnish police and prosecutors to the importance of their Singaporean captive.

Two of Eaton's operatives had preceded him to Finland, and they had assisted in confirming Perumal's identity. They had never before managed to get a visual ID on Perumal, and they wanted to make sure that this was the Raj they wanted. When Eaton arrived, he met with Jukka Lakkala and other local cops, joining them at Rovaniemi's police station, a squat building freeze-washed with snow and ice. Eaton briefed them, leading them through Perumal's history, as his evidence allowed him to outline it, where Perumal had traveled, whom he had compromised. "You have, I can assure you," Eaton told the cops over lunch that first day, "the world's most prolific known match-fixer."

The weather was bitterly cold, the Kemijoki River thick with ice. Eaton persisted in his mission, convincing Finnish authorities to align with FIFA's greater purpose. "I wanted them to realize the international impact of this guy's activity," he says. "I wanted to see them take their investigation, which on the surface looked like a simple Finnish case, to a global context." Eaton laid out all of his Perumal evidence on the table. Unlike the many international police organizations he had worked with during his time at Interpol, Eaton didn't need to be cajoled into distributing what he had.

He briefed Ron Noble at Interpol, sharing news of Perumal's arrest and an understanding of his significance. Eaton also contacted journalists across Europe, alerting them to not only Perumal's detention, but also why they should care about it. For most writers and most readers, this marked their initial exposure to Perumal and the Singapore syndicate—topics that would come to dominate sports news in the ensuing years.

Eaton's pronouncements didn't sit well with some Finnish authorities. Even though they would not have understood Perumal's importance without Eaton's evidence and urging, some felt that he had overstepped his bounds. Like police everywhere, some in Finland were concerned about who would ultimately get the credit for the capture and handling of this prolific fixer. Eaton professed not to care, and he continued to feed the Finns the information and strategic guidance that would allow them to interrogate Perumal effectively.

Over his three days in Rovaniemi, Eaton was busy. But not too occupied to consider how an encounter with Perumal might play out. After all, Perumal was just down the street from the police station, detained in the local jail. But Eaton didn't pursue a meeting, instead leaving Rovaniemi's darkness, sure that he knew Perumal as well as he needed to. "Perumal is not a smart man," Eaton believed. "Just a bold and lucky one until the latter ran out." In time, Eaton would develop a deeper interest in the subject.

CHAPTER 24

Perumal's persistent claim that he was, in fact, Raja Morgan Chelliah crumbled when the Finns sent his fingerprints to Singapore authorities and received a positive ID in return. Not only was Perumal's cover blown, but he now understood that the Singaporean government knew where he was.

When the Finnish police informed Perumal that Singapore had requested his extradition, he panicked. What could he do? Perumal had spent years making high-level political connections around the world. National soccer federations were tied into national power structures in every country. Behind bars in Finland, Perumal thought of his influential friends. If he could convince one of his political contacts to grant him citizenship in another country, this would annul his Singaporean citizenship, and perhaps the extradition request along with it.

Perumal contacted his Togo connections, applying for asylum and citizenship in their country. But he had miscalculated the fallout of the match he had arranged with the fake Togo national team. No one in Togo was willing to help. He contacted friends in Ghana, with no luck. He tried Zimbabwe. This was one of the most corrupt countries in Africa, led by the despotic Robert Mugabe for more than thirty years, a place of rule-bending and golden handshakes. Surely, Perumal calculated, the Zimbabweans knew what it meant to help a friend in jeopardy. "I am in urgent need of you guys to help me out with my present situation," Perumal wrote in an email to Jonathan Musavengana, the programs officer of the Zimbabwean national soccer federation. He continued:

I need to renounce my Singapore citizenship. In order to do that I either need to be naturalized or require permanent residence status In your country or any other country known to u. Does your country allow the naturalization of a foreigner if he is married To women in Your country. What is the duration one will have to wait from the date of marriage before he is allowed to apply to be naturalized. This is what I have in mind. I have traveled in your country in 2007/08/09. Is it possible to officially back-date the marriage with the registry of marriages in your country.

By this time, Eaton and his team of FIFA investigators were nosing around the Zimbabwean federation and its dealings with Football 4U. Eventually, the details that they uncovered would spark a major criminal case in Zimbabwe, dubbed

Asiagate. Musavengana wanted nothing to do with the man responsible for Asiagate. Musavengana forwarded Perumal's email to Henrietta Rushwaya, writing: "i even got a fone [*sic*] call from one caaled [*sic*] Aisha who claims to be his girlfriend and another one who says she z the sister!!!!!!!!!" A friend in flush times, the fiscal savior of the failing soccer associations of the world, Perumal had become a pariah.

A Finnish prison cell being more hospitable than one in Singapore, Perumal realized that he had only one option remaining. He was going to have to talk. Musonda and various RoPS teammates had broken quickly under police pressure, revealing the extent and detail of their fixing activities. The Finnish police already knew what had transpired. Presented with this information, Perumal would gain little by denying it. But the police, urged by Eaton, wanted more.

First, Perumal wanted some information of his own. He wanted to know who had turned him in. He kept thinking about his contact in Macau, to whom he owed $200,000. The man had said he knew that Perumal was traveling under a false passport. He had threatened to contact the police. Perumal thought that it must have been him.

When Perumal learned the identity of the man who had walked into the Rovaniemi precinct on February 20, he was floored. Joseph Tan Xin was a runner for the syndicate, one of the connecting lines on Perumal's international fixing diagram. And since Tan Xin had now handed off Perumal to the police, everything made awful logic.

Perumal was devastated. It was one thing for Anthony Raj Santia to use Chaibou in a match. Kurusamy had taught him that lesson a long time ago. If you weren't able to adapt to evolving circumstances, you had no business being a fixer. You had no business operating in a criminal world of deceptive allegiance. Perumal had the perfect character for such an environment. He made friendly acquaintance quickly and easily—he was a master at sincerity—though he never went deeper, and he could turn in an instant, for no apparent reason. This was the only way to get by in his world, the only way to leverage relationships for financial gain. However, if Perumal was slippery in this way, he believed that he was firm in the greater criminal code: you don't go to the police. When he discovered that Dan Tan was responsible for his incarceration, that Dan Tan had dispatched Tan Xin, he underwent a transformation. He began talking in the kinds of specifics that police and prosecutors—and Eaton—would find prescriptive in the fight against fixing.

Perumal began to outline the structure and profits of the syndicate. He claimed that the syndicate could earn a few hundred thousand euros on an operation, though its biggest scores ran all the way up to €15 million, even €20 million on a single match. Slowly, he continued to reveal information. The more he talked in Finland, the longer he would stay out of Singapore. He told Finnish police:

> The organization is structured just like a firm. There is the boss at the top, from Singapore, who decides which matches to fix, how much to pay for bribes, where to send his couriers and agents and where to place the bets. The betting mainly takes place in China. Below

the boss are six shareholders from Bulgaria, Slovenia (two), Croatia, Hungary and Singapore.

Perumal drew a crude diagram of the syndicate, which he titled "Syndicate For Fixing International Matches." A rectangular box represented each of five shareholders, the boxes connected by lines that signified runners. Along the top of the diagram, he wrote: "Winnings are tranfered [*sic*] to Spore from China through Remittance agents." Then, "Shareholders will receive percentage of the winnings even if they are not directly involved." Along the bottom of the diagram, he wrote: "Transport Cash. Analyse Games. To report the nature of the matches. To report on the performance of the players."

Beyond that, Perumal was cagey. His diagram hardly revealed much of fundamental importance. It was the traditional structure of an international criminal group, in which partners shared influence and connections, investing and profiting in established proportions. Finnish police (and Eaton) wanted names. They wanted to know the identity of the boss that Perumal kept mentioning. Repeatedly, Perumal said that the syndicate's leader was an Asian betting industry kingpin named Ah Keng. Police were skeptical, as they were with all of the information that Perumal was providing. He had spent years constructing an elaborate deception, which he peddled to players, coaches, and federations, and which had likely confused even himself, as he was the one living it.

Some of the stress that urged Perumal to his new position was artificial. The truth was that even if Singaporean officials had requested Perumal's extradition, they couldn't hope to get him. Because of Singapore's usage of capital punishment, most European countries did not maintain an extradition treaty with Singapore. Finland didn't have one. This was public information. It would not have required much research to discover it. This single fact revealed a surprising amateurism in the Singapore syndicate.

What was Dan Tan's ultimate goal in revealing Perumal to Finnish police? Dan Tan knew of Perumal's five-year prison term. This was a convenient reality. If Perumal was compelled to serve his term, he would be out of the fixing marketplace. If the goal was to initiate the extradition process that would ultimately lock up Perumal for his five-year term, Dan Tan horribly miscalculated. He didn't take into account the possibility that Finnish police might discover Perumal's match-fixing activities, and quickly, before blindly dispatching him from their jurisdiction on a passport charge. He didn't count on Chris Eaton.

Dan Tan could hardly be blamed. Soccer and law enforcement officials had never taken a proactive approach to fixing. But because of Eaton and his rapidly accumulating understanding of the syndicate, times were changing. In double-crossing Perumal, the syndicate sabotaged itself, handing police an informed source, while supplying him with the motivation to provide intricate, insider, incriminating evidence.

Perumal's arrest made news back home. Zaihan Mohamed Yusof, a prominent Singaporean journalist writing for the *New Paper,* had called Perumal the "*kelong* king." *Kelong* is a Malay word, for a fishing platform from which one casts a line. In

Singapore, "*kelong* king" had come to mean the person who rules match-fixing. In jailhouse letters to Yusof, this new *kelong* king sounded the vitriol of the high, yet real, drama that he was living. "Seeking police assistance is a violation of code number one in any criminal business," Perumal wrote. "Dan Tan broke this code and stirred the hornet's nest. And now he has to face the consequences. . . . I hold the key to the Pandora's Box. And I will not hesitate to unlock it. . . ."

CHAPTER 25

While Perumal was indulging an urge for revenge, Eaton dispatched his operatives to the Far East, which he now understood was the locus of the international fixing trade. They began cultivating sources within the Singapore syndicate, building an easygoing rapport, an approach close to a match-fixer's heart. Later, they inched closer, driving down Singapore's left-handed roads, sitting back and listening to the fixers' stories of global adventure, playing to their egos. In this way, they began to learn the identities of the main operators, began to track their movements and their emails and their phone calls, getting a grasp for how the business worked and how they might prevent it. One low-level member of the syndicate agreed to meet, and he showed up in a welder's mask, refusing to remove it throughout the interview. Eaton's operatives cultivated sources in bars and clubs and back-alley cafés. They spent time at Perumal's old haunt, the Orchard Towers, where the only thing more difficult than grasping the worth of a source was figuring out if the women on the stage in front of them might instead be men. Asia had its own way. It was no surprise that Eaton's operatives struggled to see clearly to the top of Singapore's match-fixing syndicate, to the bosses who ran it. "We know what they do and who they work with," one operative says. "We just don't know who they are."

Eventually, Eaton and his people discovered the identities of the syndicate bosses, from Dan Tan to his lesser conspirators. Behind these men were Chinese Triad groups, which controlled the betting services that had only recently come into being. "There is a two-billion-dollar weekly turnover for Asian bookmakers," Eaton says. "It's as big as Coca-Cola. And it does not produce anything. It's just paper."

The new FIFA security team discovered that in many parts of the world, all one need do to play an international friendly was show up at a stadium and pay a day's rent. There were no papers to sign, no sanctioning procedures to endure, nothing at all that made an official FIFA friendly official. There was no one asleep at the switch, for there was no switch. One operative visited the office of a match promoter behind an international friendly. The office was empty, the words "Import/Export" written on the door. When the operative called the number listed on the door, a woman answered, her voice barely audible over the crying children in the background. Another operative investigated an upcoming friendly between Turkey and the Maldives, a match listed with SBOBET, one of Asia's largest legal bookmakers. The game was set to take place at the MSN Stadium in Bukit Jalil, Malaysia. Suddenly it was switched to Petronas Stadium in Kuala Lumpur. When the agent contacted the soccer federations of Turkey and the Maldives, he discovered that neither organization knew anything about the match.

CHAPTER 26

With the data from Perumal's phone and computer in hand, Eaton dispatched two of his investigators to Sharjah, UAE, for the March 26, 2011, match between Kuwait and Jordan. When the syndicate pulled its bets in the second half, this marked FIFA's first victory against the Singapore bosses. "We put them against each other," Eaton says. "We made them suspicious of each other." As the FIFA team pressed onward, they watched the fixers professionalize. "They're now operating more like an organized crime syndicate. They're using eight, nine different phones, then throwing them away."

All turned quiet in Singapore. Dan Tan's attractive young wife answers the door of their condo, a frightened look on her face, saying that he is not at home. The devotees walk 108 times around the Hindu temple where one of Dan Tan's partners is reported to worship, but they say they do not recognize the man from a faint passport photo. At the outdoor Mon Ami Café, in the heat of Little India, where investigators say another of Dan Tan's associates pays out winnings on Monday evenings, all you can find is what's on the menu. There is a faint rustling in the apartment of Anthony Raj Santia, but the door stays closed. Chris Eaton and his FIFA security team have frightened the Singapore syndicate into apparent inactivity.

Despite his successes, Eaton realized that his approach was unsustainable. His investigators wouldn't be able to attend every suspect match in the world. Instead, he would have to institute the kinds of reforms that would fundamentally alter the way that governments and soccer administrators approached the crime of match-fixing. Eaton realized that match-fixing was "beyond policing," that he needed to take an aggressive "counterterrorism approach," that the process of prosecution and conviction would have no effect on a transnational criminal conspiracy that operated in African villages and the darkened understructure of European stadiums like any neighborhood gang. "They attract, compromise, intimidate," Eaton says. "There's always a broken kneecap somewhere along the line. This is schoolyard bully stuff on an international scale. And it's time for us to stand up and knock these bastards down."

As Eaton and his operatives progressed through their investigations in the field, some people questioned their methods. What authority did Eaton have to conduct these investigations? He wasn't a cop anymore. When Eaton and his investigators traveled to Singapore, they noticed that the local police surveilled them, shooting pictures of them with their cell phones. Eaton handed his Perumal file to Singapore police in May 2011, but the Singapore police declined to maintain communication with FIFA. Did they have to? Eaton was no longer an Interpol official.

Eaton was also encountering his first frustrations with his employer. FIFA provided Eaton with an annual budget of slightly less than $2 million for security and integrity. Meanwhile, for a VIP party in Rio de Janeiro to celebrate the upcoming World Cup, FIFA budgeted multiples of that. Eaton's irritations spilled over into his public statements. FIFA and Interpol had engaged in an agreement to establish the Anti-Corruption Training Wing in Singapore, located across from the U.S. embassy in Singapore. Ostensibly, this would educate police in the ways to confront match-fixing. But Eaton, who had experienced his own share of frustrations while dealing with Singaporean police, was underwhelmed. He said that Singapore was "a match-fixing academy." It was an apt comment, perhaps, though misplaced. "Chris said some things that most CEOs would have fired him for," says Ron Noble, of Interpol. "Because he was so direct, and because the problems reflected on his organization. Most CEOs would have said, 'cool it.'" It turned out that the FIFA brass didn't know who they had hired, that Eaton was going to address, head-on, one of their most sensitive issues.

Eaton's aggressive manner even upset the equipoise back in Lyon, as Noble privately questioned if his former charge, no longer carrying gun and badge, was overstepping his bounds as a private citizen. "The problem with Chris's approach is that he wants to investigate these cases like a law enforcement officer, but without the powers," Noble says. "He wants a FIFA witness protection program, a FIFA jump team. But it's just not possible."

Noble should have known better, for he was closer to Eaton than almost any other colleague. "Being a police officer is more than Chris's job," says Eaton's wife, Joyce. "It's who he is."

Eaton believed that traditional, long-term investigations would not defeat fixing, through trials that slogged through the courts. You had to attack the fixing syndicates like you were fighting terrorism. You had to untangle the vast web of financing that underpinned transnational fixing. Cops couldn't cross borders. But he could. "He recognized a crime of global proportion that didn't have the global attention that it required," Noble says.

Eaton began to press an internal struggle. He wrote to Valcke, urging FIFA to "respond immediately to expanding allegations of criminal international match fixing. . . . It is my strong recommendation that we cannot continue to merely respond using our administrative tools only as these challenges to the integrity of the game emerge."

Eaton contacted Michael Hershman, whom he knew from his days at Interpol. An expert in corporate governance, Hershman was the president and CEO of the Fairfax Group, a Virginia company that solves management disputes, provides digital forensics, and conducts counterterrorism operations. Hershman was more publicly identified with Transparency International, the Berlin-based watchdog nongovernmental organization that he cofounded in 1993. With Fairfax, Hershman had advised numerous companies, including GE and Siemens, on the mechanisms of monitoring internal malfeasance. On behalf of the Confederation of North, Central American and Caribbean Association Football (CONCA-

CAF), Fairfax had investigated soccer match-fixing cases in the Americas.

Over several months in 2011, Eaton and Hershman studied the ways that match-fixing had infected soccer. They discussed the proactive measures that FIFA might take to push back against the syndicate. They focused on the players, ultimately devising a plan in three parts. A whistle-blower hotline was the cornerstone of their scheme. Players, coaches, and refs possessed constantly evolving intelligence about fixing. The hotline would give them a chance to report it. The hotline would be operational twenty-four hours a day, outsourced to a specialized provider, available through email or phone, in more than 180 languages. Callers could report anonymously.

If the caller chose to reveal his identity, the next component of the program took effect. Eaton and Hershman believed it was essential to provide amnesty to players, refs, and administrators who may be involved, but who wanted to report what they knew. This was a onetime offer to escape censure from FIFA. The third plank of the program was rehabilitation, a counseling program that FIFA would offer to players who provided substantial information about fixing. In cases where the player reporting information was not involved in fixing, FIFA was even prepared to provide financial reward. A new body would oversee all of this, FIFA's Betting Integrity Investigations Task Force. On numerous occasions in the field, Eaton and his operatives had heard sources balk at providing information, citing FIFA's own reputation for internal corruption. The reforms, Hershman and Eaton hoped, would begin to spread a new message.

While Eaton drafted this reform program, he received unsettling news at home. His wife, Joyce, an Interpol employee, was bedridden. Twenty years Eaton's junior, Joyce was pregnant, and she was experiencing a series of complications. For months, the life of the fetus was in danger. Eaton worked on his reforms in Zurich during the week, traveling to Lyon only on the weekends to support his wife. Just four months into the pregnancy, French doctors discussed inducing labor. Death had only recently touched his family once again, Eaton's parents passing away in the last year. Now Eaton worried that this new life might not have a chance.

CHAPTER 27

On June 16, 2011, Eaton wrote an email to Wilson Perumal's Finnish attorney. He offered Perumal a chance to reform. Eaton wanted to enlist Perumal to speak to players, educating them about how to avoid being compromised by fixers. This was another component of his anti-fixing measures. Several days later, Perumal, restricted in jail to paper and pencil, wrote a handwritten response. He opened the letter by apologizing for the "late reply," sounding like the most mild-mannered person in the world. "I would like to thank you for considering me as a candidate to work with FIFA to help young players to stay away from corruption," he wrote. "After the completion of my trial I will be in better circumstances to comply with your request."

The letters carried the tone of a budding correspondence. Eaton was encouraged. While he pursued his investigations into international match-fixing, he thought of developing Perumal as a confidential informant. He knew from experience that there were few things as valuable, though also few things that were harder to rely on. On July 19, a Finnish court convicted Perumal of business fraud, sentencing him to two years in prison. He would serve only one.

On an August morning in 2011, Eaton was finishing a breakfast of bacon and eggs in the café of the Crowne Plaza Hotel, in Bogotá. In an effort to rehabilitate its international image after decades of domination by criminal drug lords, Colombia was hosting the FIFA U-20 World Cup. Few cities usually bid to host this event, as it required plenty of logistical effort, without much stimulus to the local economy. But the tournament was enough of a priority for FIFA that Sepp Blatter had made the trip from Zurich. Eaton had never met his ultimate boss. But there he was, Blatter, sitting at a neighboring table. Eaton introduced himself. The two men retired to a side room in the café for a private conversation.

When he was away from his handlers, which was seldom, and in an intimate setting, Blatter turned down the lights of the politician and became an interested conversation partner. Eaton was struck by his appeal. He could see that Blatter, with his suave charm, was the embodiment of FIFA's corporate culture, or what it aimed to be. Blatter listened intently as Eaton briefed him on the state of match-fixing in the game. "It's more serious than most people appreciate," Eaton said. "It needs to be strongly and quickly dealt with." Then he filled him in on the reforms that he had devised. Blatter offered Eaton all the support he required.

Eaton, the operator, was so focused on communicating the weight of match-fixing that he didn't realize that he had entered the political realm. That was where genuine concerns came to idle. What could Blatter do, in the end? Outsiders identified him as the face of the organization, but the organization was bigger than he was. Blatter looked tired. He was overwhelmed. The FIFA executive committee would soon appoint an independent governance committee to investigate evidence of internal corruption. Who knew what the committee would discover?

Blatter sat back in his chair. He sighed deeply. He assumed a look of reflection. And he spoke of João Havelange, the Brazilian soccer administrator who had preceded him in the post of FIFA president. Havelange had presided over FIFA for nearly a quarter century, from the mid-1970s, when the international game was unpolished and unpredictable, until the late 1990s, when corporate interests began to figure out how to streamline the business of soccer, turning the sport into a generator of massive profit. Havelange left the FIFA presidency in 1998, before soccer had become a global gold mine. But he could see it coming. And he understood how this great wealth would activate the thugs and connivers who had always had a hand in the game. Havelange knew that while soccer benefited from its mass global popularity, the game was ultimately ungovernable.

Blatter looked at Eaton. "When I signed the first billion-dollar TV contract, Havelange told me I was making a mistake," he said. " 'You're opening things up to predators.' "

Blatter's secretary walked into the café. It was time to head to the stadium. Blatter sat where he was, motionless, as though he hadn't heard. "I'm not a happy president," he said.

CHAPTER 28

FIFA HEADQUARTERS, ZURICH, JANUARY 2012

On January 10, 2012, Chris Eaton addressed a select group of reporters in a conference room at FIFA's Zurich headquarters. The time had come to announce the reform program that he and Hershman had developed. Sitting at the head table, Eaton read from prepared remarks: "By now, it should be absolutely clear to even the most optimistic that we are dealing with globally roaming organized crime, and that it is undermining many international sports at all levels. It is unquestionably rapacious and opportunistic."

As proof of his claims, Eaton displayed several slides on the screen behind him. These were copies of emails and letters that Perumal had sent to soccer federation officials. Eaton delineated what he called "long-term institutional answers":

Strong match and competition regulations

A clear understanding of who, in the matrix of sporting organizations, is responsible and accountable

A rigid compliance to due-diligence and sound business practices in all financial arrangements

Fit and proper persons tests, especially for match and player agents

Diligent monitoring and oversight of matches and the administration of them

A strong investigative capability when prevention protections do not work

Eaton closed his remarks by introducing the tenets and substance of the new program. The initiative would go into effect on February 1, 2012, beginning with the hanging of posters in stadium locker rooms around the world. This wasn't a solution to match-fixing. But it was the first practical step that FIFA had ever taken toward fighting it. Eaton had managed a rare feat. Since he didn't fit the FIFA culture, he was beginning to change it to fit him. He had achieved something else of surprise. While FIFA's reputation suffered from continual allegations of internal corruption, with his reforms Eaton had managed to make FIFA look like the paragon of moral integrity, if only for one day in the news cycle. With the newly appointed Independent Governance Committee casting a shadow over the organization, this press conference shone rare brightness on FIFA. Eaton had turned a negative issue into positive publicity. FIFA had finally begun to take proactive measures to combat the manipulation of its matches.

The day following the press conference, January 11, Eaton's internal office phone rang. Marco Villiger was calling.

After the two exchanged pleasantries, Villiger's tone deepened. "Chris, I have some bad news," he said. "Blatter and Valcke have decided to suspend your reforms." Eaton couldn't believe what he was hearing. They had just held the press conference. There were articles in that day's newspapers announcing the reforms. The whistle-blower hotline was set to go live in a couple of days. "We're forming the Independent Governance Committee," Villiger continued, "and this is one of the areas that they're going to refer to the committee for consideration. This is just how we have to proceed."

"You know I'm disappointed," Eaton replied. "You know how much work has gone into this. I understand the decision. But it doesn't stop me from being disappointed. Because these reforms are important."

Eaton searched for justifiable motivation in the decision of his bosses. Michael Hershman, who would go on to sit on the Independent Governance Committee, couldn't see any. "In my judgment, there was no good reason for FIFA not to have implemented a program at that time," he says. Ultimately, Eaton understood that FIFA's increasing preoccupation with its internal troubles would weaken his match-fixing mandate. Had he overreached from the beginning?

Eaton had relearned an old lesson, that even though he

might be right, politics would always dictate meaningful decisions. On the range, the lawman applied the power of the state in the way that he believed suited the environment. In society, Eaton knew that he was ultimately beholden to an organization that by its mercantile nature advanced priorities that he did not share. It didn't matter what was right. It mattered what was expedient. However much he had enjoyed the moment of his innovative successes, the moment was now over. When faced with a choice, FIFA had chosen predictably. "My concern was that outward corruption was far more endemic and far more damaging to football than internal corruption," Eaton says. "External corruption is far more dangerous to football than what happens to FIFA. Football is bigger than FIFA."

Later that day, January 11, shortly after his phone call with Villiger, Eaton drafted a letter. Addressed to Jerome Valcke, it read, in part: "Please accept this memo as notice of my resignation under the terms of my employment contract."

Eaton stayed on at FIFA for two months, handing over active investigations to his successor. The next FIFA security chief was a longtime cop named Ralf Mutschke, a German who had also been an Interpol director. Ron Noble arranged his appointment at FIFA, as he had done for Eaton. The similarities ended there. A buttoned-up company man, Mutschke focused on security rather than match-fixing. Unlike Eaton, Mutschke was an enthusiastic fan of soccer. Also unlike Eaton, Mutschke, a German, had exhibited no tendency of outdistancing his mandate.

As his days at FIFA dwindled, Eaton's thoughts began to wander. Joyce delivered a healthy baby two weeks after his resignation. The boy was named Roy, after Eaton's father. At sixty years old, Eaton was a father for the sixth time. He began to think that maybe this was the direction he should pursue, fatherhood and retirement at once, an active stasis.

He continued touring the conference circuit, dutifully sounding the horn for the fight against match-fixing, underscoring the financial scope and global nature of the problem. The influential figures of the soccer world, the grandiloquent and political, gave Eaton the cold shoulder. They looked on this Australian cowboy in their midst, this outsider, then laughed privately with one another. Eaton's hope in the possibilities of the FIFA job had been dashed, and he was slipping into irrelevance.

Eaton didn't have the powers of a cop, nor the authority of representing FIFA. What he had gained was his own independent stature. If FIFA's internal review, the battle for its corporate image, had prevented it from battling match-fixing to its full capability, then who was looking after the integrity of the game? UEFA? England's Football Association? The standard-bearers for the game had largely avoided the issue. National governments and police were focusing on domestic prosecutions. There was no one in a position of authority and influence who vigorously fought to understand the international nature of the fixing syndicates, and how to upend them.

Chris Eaton didn't even *like* soccer. And this may have been the difference. His emotions and finances and identity weren't entwined with the game's fortunes. He had no stake in the game. He had a stake in the law. While others had spent life-times in the game, and were thus prone to making allowances for its shortcomings, as though for an impolite relative, Eaton couldn't. In just two years' time, Eaton had become soccer's moral authority.

When Eaton and his corps of operatives mapped the Singapore fixing syndicate, he learned that this was only the veneer of fixing, its public face, fixing's marketing department. The power and impetus behind fixing was the manipulation of the sportsbooks, both legal and illegal, all over the world, though primarily in Asia, as Chinese organized crime overwhelmed Indonesia, Hong Kong, Malaysia, the Philippines, and anywhere they could push out the competition.

"I could see that this was no longer about sport," Eaton says. "This was about organized crime. Sport happened to be the vehicle. I saw for the first time a whole new moneymaking opportunity for transnational organized crime." He saw the unique advantage that fixing presented established criminal enterprises. When they moved drugs or prostitutes, for example, they often did so over international borders, risking detection. Fixing presented almost no risk. It was digital money, and almost entirely untraceable. "This was going to be the most dangerous income stream for organized crime in modern history," Eaton says. "This would have cashed up organized crime in a way that nothing ever had before. Also, terrorism has always copied organized crime, often copying the way they make their money. How long before they would say, 'match-fixing isn't a bad deal'?"

In March 2012, as his time at FIFA came to a close, Eaton traveled to Doha, Qatar, where initial planning was under way for the 2022 World Cup. The soccer world was just getting to know the first Middle Eastern World Cup host. Various elements in Qatari society were eager to emit all the right signals. Mohammed Hanzab, a former lieutenant colonel in the Qatari army, had recently established an organization called the International Centre for Sport Security (ICSS). Hanzab had created the ICSS as a nongovernmental, nonprofit foundation to provide expertise in the security of major sporting events. He hired a longtime German police officer, Helmut Spahn, to run the organization.

Eaton led a confidential discussion about match-fixing at an ICSS conference on March 14. At the close of the day's session, Hanzab invited Eaton to lunch at L'wzaar, a seafood restaurant in Doha's Katara cultural village. There were a few dozen people in their group, and Hanzab seated Eaton close to him for the meal. Eaton explained to Hanzab that focusing on the manipulation of the betting market would pay greater dividends than arresting players. Hanzab, wearing a spotless dishdash, listened intently. Toward the close of lunch, Eaton mentioned discreetly that he would be leaving FIFA.

Hanzab looked surprised. After lunch, he followed Eaton outside. The weather was mild, a far cry from the sweltering heat of summer. Hanzab asked Eaton: "Would you be interested in coming to work with us at the ICSS after you leave FIFA?"

Eaton thought it over. It didn't hurt that Hanzab's offer was backed by the great gas fields of Qatar. Eaton also saw a position at the ICSS as a way to finish what he started with FIFA, but this time focus his energies on the betting and financing behind fixing. He joined the ICSS as its director of integrity in May 2012, one week after leaving FIFA. As he selected his staff, events in Finland began to play out.

CHAPTER 29

Wilson Perumal was nearing the end of his prison term. By standard legal mechanism, Finnish authorities would have deported him to his previous port of embarkation. That was the United Kingdom, which was one of the only European countries that maintained an extradition treaty with Singapore. Perumal might have traveled this route to his five-year prison term, though it's doubtful that Dan Tan had conceived of it. Faced with few options, Perumal cut a deal.

Under the guise of Europol, the European Union's criminal intelligence body, Perumal agreed, upon his release from Finnish prison, to be transferred to Hungary. He had deep associations in Hungarian organized crime. He often traveled with Hungarian partners in order to carry out fixes. Authorities in Budapest were in the midst of a wide-ranging investigation that eventually implicated several hundred figures in domestic soccer leagues. Police and prosecutors from Italy, Germany,

France, and other European countries would eventually visit Budapest to interview Perumal, who, Eaton heard, was living under close monitoring and protection. It looked like Perumal was going to be in Europe for a while.

Eaton was gone from FIFA, but the issue of match-fixing had a hold on the organization. The month following Eaton's departure, FIFA held its sixty-second congress, in Budapest itself. Blatter and Valcke led the festivities. Ron Noble was there. They reviewed organizational priorities. They retired to the cafés along the Danube, which separates the city's two halves. And who among their party understood that the man who most threatened their security and longevity was just a few city blocks away, talking to Europol?

While the sporting and law enforcement brass felt safe in the approaching summer breeze, Perumal was in jeopardy, as he shared information with European authorities. Back in Singapore, underworld figures approached his old associate, Dany Jay Prakesh. "Protective custody," Prakesh said. "They better have him locked away real tight right now. They were looking for him in Hungarian prison. They offered to pay me three hundred thousand dollars to go to Hungary just to help them identify him. They want me to sit on a roof with the sniper. It's easy. Fifteen minutes alone, he walks to the bathroom, he's dead. The Russians made three hundred and fifty million dollars last year on this. The Italians, the Bulgarians, the Hungarians, the Croatians. Wilson is messing with people's business."

CHAPTER 30

DOHA, QATAR, 2013

On the veranda of the Wahm Bar at the W Doha Hotel in Qatar, Chris Eaton raises his beer in his right hand. "A steak in every glass," he says, justifying the nourishment. He toasts the four men standing around him. These faces are familiar, because they are all the same as before. Eaton's FIFA investigative team has transferred with him to the ICSS.

A searing Arabian breeze sculpts the terrace. All around Eaton's group stand the architectural markers of sudden wealth, along with the ambition and impatience that this stimulates. Doha is an architect's palette. Anything goes here, any shape, any material, any color. One tower looks like a bottle of perfume. Another skyscraper appears to be layered in white gold. There is nothing within sight that is in keeping with nature's sandy surroundings, nor is there harmony between these structures. This place looks like a collection of lost towers from other places.

This is where the World Cup will take place in 2022, on the Qatari peninsula, which juts, aridly, one hundred miles into the Persian Gulf, Saudi Arabia the only bordering neighbor. Viewed on a map, Qatar looks like a tick on the backside of Saudi Arabia. But Qatar is no bloodsucker, with the third-largest natural gas reserves in the world. There is virtually no unemployment here. Nor are there any taxes.

About one-eighth of Qatar's population are citizens of Qatar. The rest of the people within these borders are here to serve, engaging in the remittance economy. There are more Indians here than Qataris. There are more Nepalese. On an afternoon in June or July, when temperatures in Doha reach 135 degrees (as soccer fans now know), from the comfort of an air-cooled sedan you may witness these foreign laborers fulfilling their duties on outdoor construction sites, somehow surviving beneath the sun. The towers must rise, and now the stadiums, too.

It may be difficult to accept that an organization that is meant to uphold the integrity and transparency of sports, the ICSS, comes from a country where political parties are outlawed, where there is no elected legislature, where a woman's testimony is worth little in court, where one family, the al-Thani, has reigned since well before the Cambridge Rules codified soccer, at Trinity College, in 1848. In the vacuum of soccer's leadership, however, those who might have led relinquish their right to complain. Ennobling burdens fall to those eager to assume them.

Shortly after beginning with the ICSS, in May 2012, Eaton encountered the question of how a man of professional integrity such as himself could work under questionable assignment. Conventional wisdom led most people to believe that the Qataris were corrupt, repressive, backward, and that they had used bribery to secure the World Cup. Eaton took to high ground, stating that he "peddled facts, not allegations." He could produce an affidavit from a Qatari whistle-blower who had initiated the World Cup scandal; the woman wrote that she had fabricated the entire ordeal. "Look, I was the head of security at FIFA when those allegations emerged," Eaton would tell those who asked. "I never saw any evidence. In my assessment, the Qataris were perfectly clean." His years at Interpol had granted him professional experience with the cultures of the world, and he did not carry prejudice into a region that inexperienced Westerners assumed operated under deep deceitfulness.

Nevertheless, the questions unsettled Eaton. He learned to accept this collective doubt about Qatar and himself as the price of opportunity. Considering the fact that FIFA had collapsed his ambitions, who else was hiring? "The ICSS is my opportunity to finish what I started," he told himself. Eaton is not the only Westerner occupying a senior position at the ICSS. Just as the Qatari government owns Al Jazeera, the influential international TV news network, so the ICSS would utilize domestic funding to coalesce international expertise, creating a new utility for the world. At least, that was the idea.

The ICSS offices occupy the thirty-third floor of the Al-fardan Towers, in Doha's West Bay. At sixty-four stories, the towers are the tallest buildings in Qatar. They are modest steel and glass structures, built as residential buildings. Eaton's office is a converted bedroom with a large desk that he does not use, instead taking meetings in an extended lounge. From his window, he sees the city in a sweeping view, looking toward the Pearl.

This is where he lives, having left Zurich behind. The Pearl is a thousand-acre artificial island built on the former site of Qatar's main diving area, pearling having been the heart of the country's pre-oil economy. The island was designed in Venetian style, with interlocking canals. Waterside villas are painted in bright blues, reds, yellows. It is a beautiful place, though lonely, with low occupancy. Eaton resides in a three-story villa with Joyce and Roy. Their sector on the Pearl, the Qanat Quartier, is so devoid of residents that the stray Jet-Ski that cuts up the canal is the only reminder that the Eatons are not alone out here in the aquatic desert.

A Venetian ghost town, the Pearl was a quiet place for Eaton to reformulate his approach to the match-fixing fight. He was now freed of FIFA's institutional restraints. He didn't have to protect the image of soccer's corporate body at the expense of the game's integrity. And his horizon now went beyond soccer. Eaton's position at the ICSS required him to investigate all sports, since all sports were afflicted with manipulation, to varying degrees. Horse racing, rugby, cricket, tennis, volleyball, baseball. "You know where it's big?" Eaton asks, rhetorically. "Badminton. Huge."

The particular sport didn't matter, because Eaton was no longer investigating matches and fixers. He was now interested in the people who controlled the fixers, those who owned and manipulated the sportsbooks. Fixing in all sports emanated and terminated in the same place. Ultimately, the money flowed in and out of Southeast Asian organized crime. To defeat fixing, Eaton and his investigators would have to learn how Asian and European criminals cooperated in betting manipulation. "What is the opportunity that is causing this attraction by organized crime?" Eaton asked himself. "Betting fraud. The real target should be the criminals who handle the betting fraud that inspires and funds match-fixing. It's like international banking in the 1960s and '70s. It's a free-for-all in Asian betting. I want to know their MO. I want to know who they are. We're going to get to the heart of organized crime's interest in match-fixing."

As Eaton outlined his strategy to himself and to Hanzab and to underlings, he heard news of an old contact. Through his police sources in Europe, Eaton learned that Wilson Perumal was not being held in detention in Budapest, nor in a safe house, though his life was in danger. Nor, it appeared, were Perumal's activities particularly restricted. "I heard that he was living it up in Budapest in the nightclubs, having a great time," Eaton says. "The injustice of it."

As European police used the information that Perumal gave them to arrest players and referees and soccer administrators, Eaton moreover didn't understand how this approach might impact the greater fixing syndicate. Though operationally significant, players were ultimately incidental, interchangeable, renewable. The only way to strike at fixing in a meaningful way was to map the criminal structure and attack those at the head of it. This required a coordinated strategy developed and carried out by police and government authorities in multiple countries. But as in Finland, authorities in Europe were using Perumal's information to arrest those people who operated locally and could be charged locally. The ambition and tenacity of European prosecutors and judges elapsed at their national borders, the same borders that were meaningless to the syndicate.

"I didn't see a global strategy behind it all," Eaton says. "I saw this guy, this global fixer, laughing his head off. We're looking at someone who was clearly the glue that brought it all together. And he's the only one who doesn't get the maximum penalty. It struck me as fantastic."

Frustrating though the situation was, Eaton could do little to influence it. He furthermore had his own easterly operations to construct and manage. He dispatched his operatives to Southeast Asia, providing them with a single guiding principle: "Even with all the Sportradars in the world, you will not know what the criminal is doing. There's only one way you will know what the criminal is doing. And that is by speaking with him."

CHAPTER 31

In July 2012, Perumal learned that a Swiss court had implicated João Havelange, the former FIFA president, in a bribery scandal. The court claimed that Havelange had pocketed $1.53 million through a bankrupt marketing company that had once held contracts with FIFA and the International Olympic Committee. His alleged partner in the scheme, Ricardo Teixeira, his former son-in-law, was the head of Brazil's 2014 World Cup organizing committee. Teixeira resigned his post. Perumal wrote an email, asking Zaihan Mohamed Yusof, from his hometown the *New Paper,* to forward it to Eaton:

> FIFA looks more like a criminal enterprise than a official organization. The pot calling the kettle black. These guys are criminal who deserve to be sentenced to serve prison terms and not judged by an ethics committee. A very convenient way to escape

from justice. A bunch of crooks trying to go around
the world to preach against corruption. FIFA is a joke.

FIFA's behavior was so questionable that it allowed a convicted match-fixer to appear credible in his criticism of it. It was
hard to mount an argument against Perumal's claims. Eaton
realized this. Recognizing an opportunity to revive direct communication, he wrote to Perumal:

Wilson I do not admire what you have done in football,
whatever your motivation and circumstances. And I
will not try to humor you like I am sure others have
done to try to get you to talk. But I do like what you
are doing now. It will help football, indeed all global
sport, to face up to the realities of the global greed
and it's-just-a-business culture that has corrupted
so many. Being real and facing up to the ugly past in
global sport is urgently needed today, and directly and
indirectly you are playing your part in that.

Now I well know there are many others who have
done far more damage than you, many others. And
if they were showing the backbone you are, I would
be asking them too. But most are just shadowy,
greedy bastards without consciences. You certainly
were not shadowy, and that might have been your
undoing, and you are showing a conscience—good
for you.

I left FIFA after two short years for many reasons
Wilson, but mostly because I saw an opportunity to

work across all sports. To the extent i have any ability to do this, i want the shame that has hit football to be prevented in other sports, and the resistance of those who are vulnerable to corruption, strengthened. If you think I left FIFA for money—well it is true that you don't know me. If you did I am sure you would have another opinion.

I am working as hard as ever and with some excellent people, to bring integrity to international sport where it needs it, and it does need it. Something you know better than most.

Keep safe Wilson, and I will be happy to meet you some day and try to understand you more, and perhaps you to understand me more.

Though Perumal may not have sensed it, Eaton was trying to cultivate him to eventual use. Eaton knew that this would take a while. Meanwhile, Eaton maintained close contact with his ICSS investigators in Southeast Asia. The time had come to manufacture new results for a new employer, one that was eager to show the world that it deserved its World Cup.

CHAPTER 32

The bookie was late. No surprise. These are not people who operate by any clock other than the one that counts down the time remaining in a match. Jakarta's street buzzed beyond the door of the café, the constant stream of motorcycles and scooters navigating the traffic. Indonesia was a logical locus for two of Eaton's ICSS operatives. They had arrived here after visits to the Philippines, Hong Kong, Macau, Vietnam, and Singapore. Their investigation into bookies was under way. And while they waited for this bookie, a local, ethnically Chinese man, they discussed the seriousness of the environment. Players in Indonesian leagues had met with bad ends, or such was the gossip. "If you stop the fix here," says one of Eaton's men, "they'll kill you."

Sports betting is illegal in Indonesia. But Indonesians of Chinese descent established an illegal gambling market here decades ago, in the days before live betting, and before the Chinese economic boom. At the time, many of Indonesia's 250 million people possessed the expendable cash that would soon come to the Chinese. The market was controlled and managed by the ruthless Triad criminal groups that underpinned all of Chinese vice.

Triads financed all bookmaking enterprises, opening them, closing them, relocating them, renaming them in rapid succession. The operators of the books assumed great responsibility. A bookie might have to take a fall with the police, because the syndicate said so. Sometimes a bookie had to accept a bet on a fixed match, understanding that he would absorb a loss, perhaps one that would put him out of business. Why? "Sometimes to stay alive," says one of Eaton's operatives. "This is not a rational market."

Eaton's investigators discovered that over time, organized crime had developed algorithms similar to Sportradar's, writing code to monitor the betting market for weaknesses. They watched how the money flowed, timing when to place their bets. "Sportradar is a big company, with big turnover," says one of Eaton's investigators. "But in organized crime terms, they're not even on the map. If you want to throw forty million dollars at it, organized crime will throw four hundred million. Everything's relative to the amount of money that you have and the amount of expertise that you can buy." As Sportradar and others pushed the market toward sophistication, the syndicate outdistanced them, evolving its technology for speed, to be able to place bets as quickly as possible, their models the speed traders of the financial markets.

Eaton's men learned that the syndicate would put together a betting team for big fixes, a chance to put massive money into the market. Speed was essential. They had devised a system by which one worker could place 127 bets in three seconds, the wagers varying from $500 to $5,000. Timing was as critical. If the betting team flooded the market for a red card in the eighty-fifth minute, laying thousands of bets in three seconds, the bookmaker wouldn't be able to react. Sportradar's algorithm would pick up the inconsistency, but too late for anyone to do anything about it.

Sources talked about Russian organized crime and the way it was evolving. Russian criminal leaders had started recruiting indigenous criminals in various countries, providing them with their organization's full backing—financing, drugs, women, weapons—marrying foreign elements into a previously closed brotherhood. By fostering this transnational unity, the Russians had expanded their reach globally, rivaling the Triads for supremacy. The bosses in the betting syndicates were always either Chinese or Russian. Everybody respected the South Koreans for their rigid criminal discipline. A name that frequently surfaced was Dawood Ibrahim, India's most powerful criminal, whose so-called D Company syndicate ruled the fixing of cricket in Bangladesh, India, and Pakistan. Government immigration files revealed that Singapore's Tamil fixers often visited India, and Eaton's men found it of professional obligation to consider this something other than a coincidence. They traveled to Manila to investigate the advantages of lax government regulation. "Once you have a legitimate front, you can approach legitimate business, as well as take money from organized crime," says one of Eaton's investigators. "There are a lot of companies out there that are backed by the Triads. You'll never find out who's actually behind them."

They traveled to Macau, the casino capital of the world, a choppy hour-long ferry ride from Hong Kong. They examined the Wynn casino, walked the main floor of the Sands. They saw how this differed from Las Vegas. Nobody was drinking. The low limits looked like high limits, fifty dollars the minimum. There was no energy in the room. This wasn't entertainment. This was gambling, though it wasn't the most serious gambling of all. On an upper level, behind frosted glass, that's where the junkets played out.

Triads controlled the junkets. A Chinese gambler would deposit cash into a Triad account on the mainland. Then he would join a small group of high rollers for the trip to Macau. The Macau casinos counted on these whales for 70 percent of their bottom line. Behind the closed doors of the junket, the casino answered all needs and urges, and the gambling was constant. The casino would give the gambler a stack of chips, his stake against the cash he had deposited on the mainland. At the end of the night, or at the end of the weekend, or the week, if the gambler was in the black, the casino would cut him a check for the total. If he was in the red, the Triads would kindly inform him to cover arrears.

For the match-fixing syndicate, the junket was the endpoint of the field activities of Wilson Perumal. Junkets at Macau casinos were a revolving laundry. When the criminal used the fix to win on illegal Chinese sportsbooks, he had to find a way to make that money legal. He would transfer it to a mainland Triad account, then head to Macau on a junket. With chips in hand, he wouldn't even have to gamble. But if he wanted to fit in, he could mark out a place at the roulette wheel. He could place half of his chips on red, the other half on black. When the ball found its slot, the casino paid out, taking its cut. Then he could cash out, receiving his clean casino check from a legitimate casino account.

The only way to get in was to dive in. Eaton's ICSS operatives were passing themselves off as gamblers looking for a fix. They had $25,000 to throw around. The plan was to get advance warning of a fix from the syndicate contacts they had made while working for FIFA. With that information, they would place their $25,000 on the proper match, and in the proper way. If they did that a few times, they would build their stake. Eventually they would have the kind of cash that would get them in front of an influential bookie, not the one who finally arrived at the Jakarta café, looking like he was in the middle of a weeklong bender.

CHAPTER 33

On June 1, 2012, Rushwaya received an email from Perumal. "I heard you went through quite a lot of trouble," Perumal wrote. "Don't worry vengeance is a dish best served when it is cold. . . . Fuck FIFA, Fuck the ethics committee. How many trusted men do you still have in the team. They play Guinea. What is the strength of the team. Kindly send me some details." Rushwaya shared the email with Eaton.

Eaton discussed it with one of his investigators. "These are coppers behind this," Eaton wrote, determining that the email was "a fish hook," a device meant to ensnare Rushwaya, involving her in a new fix. Zimbabwe would face Guinea only two days after the email had been sent. Eaton believed that Perumal's Europol handlers in Budapest were trying to manipulate Rushwaya through false sympathy for the troubles she had recently endured. He was underwhelmed by the tactics employed by "these Europol twits."

Eaton recognized a pattern. He knew that Perumal had

begun reaching out to his old contacts. Perumal was writing emails to people in Singapore, to referees he had used in fixes. He had resumed posting to Facebook, reviving his dormant network. Eaton surmised that it was a ploy, and a clumsy one, to engender further arrests and investigations. And he wanted in. Eaton wrote to his investigator: "Let's have some fun with all these pricks—criminal, coppers and football fucks—if we can't beat 'em, let's fuck with them."

Eaton advised Rushwaya to reply to Perumal. On June 6, she wrote:

> Hi Wilson
> I am really surprised and happy to hear from you. Yes times have been hard but I think they have been
> Hard for you too. I will be happy to see people get what they deserve.
> The papers say you are in prison. So where are you now? Can we meet in Africa soon to discuss? I am still close with many of the team. The keepers, centre back, left and right back, and two mid field-ers. What do you suggest for them?
> We lost 1–0 to Guinea at home. Sorry for the late reply. . . . We are playing Mozambique this weekend in Maputo. What do you suggest?

Perumal answered quickly:

> Are you very sure the players can work. i don't want anymore trouble for anyone. I think it is a loosing [sic]

battle in Maputo. But the players have to think likewise. If you can get the mobile numbers of the players
in Maputo we can work. Since you are going to be in
Maputo as well then you can arrange the rest.

I have a man in Maputo. Please pass me your
Mozambique number as soon as you get one.

I will call u in yr zim number later today.

Bye. Best regards.

No one knows i am out. keep it that way. i dont
trust anyone other than u.

The correspondence eventually dissipated, and it left Eaton
wondering just what Perumal or his handlers were trying to
achieve. He had told his underlings that the only way to know
what a criminal is doing is by speaking with him. In late 2012,
Eaton emailed Perumal. Eaton wanted to maintain contact
with Perumal, simply as a point of operational utility. He was
also piecing together details about possible fixed matches in the
World Cup in South Africa.

Eaton had received information from South African Customs. He knew that Perumal, traveling on the Raja Morgan
Chelliah passport, had been in South Africa for almost the entirety of the World Cup. Eaton had other evidence of Perumal's
presence in the country during that time, including photos that
had been posted to Facebook. He didn't let on. He asked Perumal
who may have been in South Africa at the time, who may have
been in a position to manipulate matches.

"Please note that my passport was impounded by the Singapore police during this time," Perumal wrote to Eaton. "Obvi-

ously i was not in a position to travel." He advised Eaton to look into Santia's activities. He claimed that Dan Tan was staying at the Garden Court hotel during South Africa's matches against Colombia and Guatemala. "Wilson Raj was in Singapore busy with his case during this period," he wrote. "I may have been the CEO of Football 4 U that does not mean i rigged all these matches."

Eaton replied, hinting of his knowledge of Perumal's emails to South African officials. "I understand you have to be careful, and I want to help if I can to reduce your personal exposure in SA, which will become explosive, be sure of that," Eaton wrote. "So give me something I can use on these guys. I think you are too important on many other issues that I hope we can work on in the future."

Perumal stonewalled him. He continued to point the finger at Santia and Dan Tan, claiming that he himself hadn't been there, wasn't involved. This approach was trying Eaton's patience. He responded:

> It's very important that we trust each other as far as is possible. . . . No matter what is happening in Europe, international political pressure driven by national embarrassment will be very strong indeed to find a scapegoat.
>
> You need to very seriously consider you position here Wilson. In my opinion, which should count for something to you, you will need to do more than what you have so far disclosed in Europe. . . .
>
> This is my candid advice to you. I'm interested

in making the best of this emerging storm for you so that we can really clean up football.

February 4, 2013, was a signal day in the fight against match-fixing. This was the day that alerted the masses. In The Hague, Netherlands, Europol convened a press conference. Rob Wainwright, the director of Europol, announced the findings of Operation VETO, an eighteen-month investigation into match-fixing on the continent. Wainwright announced that Europol had uncovered nearly four hundred manipulated matches in fifteen countries, involved 425 players, refs, administrators, and syndicate criminals. Wainwright singled out activity in leagues in Germany, Switzerland, and Turkey, as well as related matches in Africa, Asia, and Central and South America. He intimated that the fixing syndicate had manipulated a UEFA Champions League match, as well as two World Cup qualifiers.

For those with knowledge of the issue, there was little revelatory information in Wainwright's announcement. Most of the matches that Wainwright discussed were part of ongoing criminal cases. They weren't news. What the press conference did reveal was the underlying tension between Europol and Interpol, as the two organizations, while working closely on many issues and cases, fought for credit and publicity. By grandstanding with information that was already in the public domain, Wainwright was trying to take law enforcement's lead position on match-fixing.

Amid the political infighting and the petty arrests and the unending investigations and trials, match-fixing continued on every continent. What was being done to combat it in a meaningful way? It had been more than a year since Italian authorities had issued an arrest warrant for Dan Tan. And while Dan Tan had spoken voluntarily with Singaporean authorities, he was still a free man. Wainwright had failed to mention that his star informant, Wilson Perumal, whose evidence had contributed significantly to Operation VETO, was doing all he could to put Dan Tan in the Singapore jail cell he himself knew all too well.

CHAPTER 34

What a feeling, free in Budapest. The Danube sparkles on a summer evening. The night-light Buda Castle reflects in the wrinkling waters, a spectacle of middle European gentility. The city's squares are alive with chatter, lager, leisure. The Chain Bridge stands ready to convey you across the Danube in nineteenth-century grandeur, and what a pleasant way that must be to move. Here come the professionals. On the esplanade, these girls aren't playing games. There are too many of them, competition forcing them to be more straightforward than maybe you are ready for them to be. In front of the InterContinental hotel, they approach one by one, curious about your evening plans. They form groups, then packs, until it appears that the Hungarian capital is made for them and no one else. That is Budapest, beautiful tragically, unburdened by the petty morality of other places, ancient, permissively inclined to whatever you got.

Should you feel like gambling, go ahead, for here it's understood that life itself is such, a proposition. The Las Vegas Casino is not a poor place to do it, located here on the Pest side of things, there on the ground floor of the Sofitel hotel, off József Attila Street, paces from the Danube. The taxis whip by the casino, heading away from the water and on to inland adventure. These cars and girls for hire confuse the summer street. So do the people ambling about drunkenly or sullenly in search of something to do, the sun now gone down. There is the one person standing still amid life's proposition, the red Las Vegas lights emblazoning his features.

Wilson Raj Perumal asks: "How about a bite to eat?" He stands six feet tall, and he fills a red rugby shirt with solidity, though he shakes hands weakly. Perumal is forty-seven years old. His hair is graying at the sides. This is the only visible sign of advancing age. Otherwise, his eyes sparkle. Why shouldn't they? Summer in Budapest feels like a reward.

The restaurant is called Spoon. It is located on a boat, moored to the Pest embankment. It's a date place, with attentive service. Sitting at the table, ordering a snack, Perumal is amiable, polite, interested, alert. He speaks English in the characteristically Indian accent, and for some reason that is quaint and kindly to hear, as though the speaker were incapable of doing any wrong. He asks, "Is this your first time to Budapest?" as though he were a local.

Shortly after Perumal arrived in Budapest, in early 2012,

police arrested a few dozen players in the Hungarian league, most of them from Rákospalotai EAC, a club located in northeastern Budapest. Local sources claim that these were low-level figures, that the government doesn't have the will to target the leaders of Hungary's match-fixing ring. Perumal is due to testify in an upcoming trial. He has said enough to earn a series of conditional one-year visas, expiring and renewable each May. But he exists here at the whim of the federal prosecutor's office. "At the moment, they are finished with me," he says. "They might fuck me up, because this is Hungary. I don't know what they might hand out next May."

While the local authorities have provided Perumal with no guarantees, they have allowed their guest remarkable latitude. "The first year I was here," he says, "I went to the clubs almost every night." He smiles at the memories. "Hungarian women are very friendly." Things have since changed for Perumal, he claims, the thrills having run their course. He says that he now lives with a woman, a local, who is twenty-two years old. They have been trying to conceive a child, though with difficulty. "We're going to try in vitro fertilization," he says. He has heard that Hungarian law allows for the father of a child locally born to petition for citizenship after a few years. Perumal may have engineered a way to avoid Singaporean prison permanently.

In the meantime, while in stasis in Budapest, disconnected from the fixing network that he developed over the years, Perumal says that he has designs on a new line of work. "I'm going to open a little restaurant. An Indian place. You can't find a good one here in Budapest. Part of it will be for gambling, with a few terminals where you can bet online." This is how Perumal claims that he earns his keep for now, placing wagers on soccer matches through his standing account at the IBCbet portal. "If I am careful, and I don't get carried away, I can earn between one thousand and five thousand dollars per day," he says. "If I'm not careful, I can lose a lot more." He appears reconciled to his new circumstances, lucrative projects and transnational criminal conspiracy now fragments of the past. The old associations have gone by also. "Nobody wants anything to do with me," he says. "I'm too hot right now."

Spoon's many empty tables make the diner feel unwanted and alone. Perumal's characterization of his life in Budapest is no different. He says that the locals are difficult to befriend, or at least difficult for him, though he doesn't much mind. "I have only two friends in the whole world, people I could trust with a million dollars. I don't need more."

Would two people be enough to kick-start the old business, maybe run a few fixes remotely? Perumal says that he feels no temptation to do so. "I don't want to jeopardize my standing with the local authorities," he says. "Anyway, that's all over for me. I'm stuck here. My wings are clipped."

Perumal leans back in his chair. He lays his hands in his lap. He looks through the windows of the restaurant, across the water to the Buda Castle. "I figure I have only fifteen, twenty more years to live," he says. "I can hang here."

The night pushes on to a place called Romkert, an outdoor nightclub on the opposite side of the river. Here is the usual convergence of vaguely available young women and guys in tight T-shirts with indecipherable writing silk-screened onto them. The music is loud, though the energy is low. Perumal says that Romkert is usually livelier. "Sometimes you gamble and you lose," he says, leaning an elbow against the bar. He orders a drink.

It doesn't take long to realize that Perumal is the sort of person who feels uncomfortable during the natural lulls of a conversation, is compelled to fill them. Among his preferred topics is the Singaporean justice system, and how it has wronged him. "I was sentenced to five years for nothing. I am so angry, sometimes I cannot go to sleep." Five years for sparring with a security guard is an exorbitant, typically Singaporean sentence, and this legal episode would logically preoccupy the person at the unfortunate end of it. But while Perumal claims innocence of a sort concerning his airport charge, he is surprisingly forthcoming about his fixing schemes. He talks about them in great detail and wonder, as though he, in his new pedestrianism, finds these past experiences hard to believe. To him, they are hardly crimes at all.

"When I started branching out, it was so easy," he says. "I realized that Bahrain had no money. They said, 'Wilson, so long as you bring in a team, you pay the airfare, you pay the hotel, we are prepared to play. We're more than happy. Thank you very much.' Same thing, when you go down there with the referee exchange program. 'We take your referee, throw him somewhere. We send you one. Our referee is going to do something dirty. Your referee don't have to do anything dirty.' This is how you speak. They're happy. Who wants to say no? A lot of them, they don't have any money. FIFA is not giving them enough money." He takes a sip of his drink, then continues. His energy level has risen. "Football has got no money. There is money in Europe in the highest leagues. If you go to division two, it's already no more money. In Portugal, in Spain, they're barely making ends meet. Match-fixing is not gonna end."

The subject that elicits the most passion from Perumal is FIFA. The organization's troubles with corruption make it an easy target, though a worthy one, even if Perumal may be engaging in a game of misdirection. "FIFA is like a mafia organization," he says. "They don't answer to anybody. They don't give a fuck about anybody. What I can't digest is who would want Qatar to host the World Cup in June? I've been there in June. You can't even walk around in a T-shirt. Something is very wrong. Even a blind man would not make this mistake."

Perumal contrasts himself to the body charged with overseeing the welfare of the game. "Fixing doesn't touch anybody," he says. "There is no victim. I have turned people's lives around. African players would go home from a tournament with ten thousand, fourteen thousand dollars in their pockets. Nineteen years old. That kind of money changes their lives. I have helped players get operations for sick people in their families. I have put roofs over people's heads." And he has provided for the lucrative Nigerien retirement of his favored FIFA referee. Perumal claims that he paid Ibrahim Chaibou roughly $500,000 over the course of their relationship. "Chaibou is now living with his four wives in Niger. He lives like a king."

A group of four women gather near the bar. They are chatting closely with one another. One of them trades looks with a guy across the dance floor. There are many shadows in this nightclub, and many more in Budapest. Perumal knows that the more he divulges to investigators, the more enemies he makes. Yet he claims that life's dangers and dark pathways do not preoccupy him. He prefers not to hide. "I'm not afraid," he says, looking around the club. "I have seen a lot. I live my life. It's better to be in the thick of the action."

He leans toward the bar. He waves over the bartender and orders a bottle of champagne. As he waits, Perumal expands on his words. He knows how easy it is to get to someone, because he was the one who for years got to people. "It's always possible," he says. "Human beings are vulnerable. We are all vulnerable."

The bartender pops the cork of the bottle of champagne. He passes the bottle over the bar to Perumal, along with a few glasses. With these items in hand, Perumal approaches the group of four women in front of him. It's a gamble. "Ladies," he says, grinning, "would you like some champagne?" Wordlessly, the girls disperse.

CHAPTER 35

Out of a taxi, and walking toward another nightclub, Perumal elucidates a tested strategy. "You take good, first-division players, and you move them down to a second-division club. They're so much better than everybody else. It's like a horse race. You tell your jockey to hold the horse until the last part of the race. Then he lets him run. You can control the race." When Perumal explains the ease with which he executed his fixes, he does so with obvious regret, feeling as though he left money on the table. "The Chinese betting services only came around in 2010," he says. "If they had been there before, I would have made big money. I had matches in 2009 and nowhere to take them." He says that he earned roughly $6 million from fixing. "If I hadn't gambled, I would have made twenty-five million dollars, easy."

You don't need big money to hang out at Morrison's 2. This is a one-room dance floor, dark, located on the inside of a Budapest courtyard. Perumal pauses in the entryway. "I used to come here every night." He shakes hands with the bartenders. They greet him by name, ask him where he has been. Over the next hour, several people materialize from the dance floor and give Perumal a hug, a handshake. He buys the drinks. A smile is affixed to his face. He is comfortable here, a club guy. It is not difficult to understand how an impressionable soccer player, maybe far from home, maybe young, maybe looking for a good time but unsure of how to have one, would find comfort under the wing of someone so fluent with the night and hard currency. This is no Indian restaurant owner. This is the *kelong* king, shackled for now in Budapest. Perumal strays from the bar and toward the action of the dance floor.

Later on, after Budapest's roads have emptied, Perumal takes a ride. The Internet café has only a few other clients. They're playing games. Wilson Perumal is betting them. He enters a simple URL: www.ibcbet.com. He types in his username and password, lengthy combinations of letters and digits. His IBCbet account appears onscreen. He scrolls through the many leagues matches, friendlies, and youth games that the book is offering for betting. In the café's flickering light, Perumal gambles.

I'm a gambler," Perumal says. "If I hadn't got caught in Finland, I would have gotten caught some other time." It is the next night in Budapest, and strange objects hover in the sky over Vörösmarty Square. For the tourists at the cafés. Street vendors fling the objects into the air, twenty-five feet high.

They are made of nylon and taffeta, small enough to fit in the palm of your hand. They are light, and they flutter to earth in a whirling motion, lights blinking. Small children, street children, dart around the square, trying to get their hands on one.

The action fails to distract Perumal, who sits at a café table, coolly sipping a Coke. "Chris Eaton doesn't have a good reputation among the European police force," he says. "He's a little too loud. The German police despise him. He's not professional."

A street vendor strays near Perumal. The man shoots several spinners into the sky over the plaza. Then he moves on a few paces, courting customers. He is a freelancer, an entrepreneur, drumming up his business. Perumal is suddenly full of ideas.

"You know, you could take the Cameroon domestic team, and they would play as the national team abroad," he says. "I could arrange to have Egypt play against Colombia in the U.S. You play in Texas or California, and you will sell out the stadium, with all the Mexicans." He is energized at the thought of the action. He has been stuck here for a while now, watched and restricted and undergoing withdrawal, recalling what it used to be like to be in the mix. "A gambler takes risks. When you have no more chips to push, you find a way. You have to lie. Gamblers are like drug addicts. We will bullshit. We will take ten dollars out from your pocket. Because this is an addiction."

As though the prospect of feeding the addiction has subsided, Perumal's eyes cast toward the cobblestones of the square. He realizes his predicament. When he speaks again, he sounds resigned to it. "Maybe I'm not good for football. I'm persona non grata in football circles. Undesirable person. I'm not a well-liked person in football. Maybe people think

I'm damaging football." He pauses. "Anyway, I'm out of all that now. My wings are clipped."

One of the spinners drops from the sky. It flutters erratically, left and right, its pink lights flashing, before it lands at Perumal's feet. He looks at it for a moment, contemplating this lifeless, worthless object. He looks away, disinterested.

Someone approaches. It is a little gypsy girl. She is not more than three feet tall. Something about her isn't right. She looks malnourished.

The girl reaches for the spinner. But Perumal is too fast. He leans forward in his chair and snatches it up before she can grab it. The girl wants it, which means that it now has value. Perumal holds the spinner in his hand. He twists it around in his fingers, the glow lighting up his eyes.

The girl holds out her hand. Perumal offers her the spinner. She reaches for it. But he pulls away his hand, and she grabs only at the air. This confuses the girl. Her features are twisted up, as she tries to understand what's happening.

Perumal holds the spinner over her head. The girl jumps for it, but he lifts it just out of her reach. The girl's mood changes suddenly. Now she is angry. But when he goads her into another try, dangling the spinner above her head, the girl jumps again. She glares at him. She stomps one of her feet on the square. Perumal relents. He lets her have the toy. His face shows no thought or emotion.

Now that the girl has her spinner, she turns away from Perumal. She takes a few steps. Then she thinks better of it. She stops. She turns around. The girl spits at him.

CHAPTER 36

HONG KONG, PRESENT DAY

Oscar Brodkin has come to Hong Kong by way of London. He characterizes the transit as "FILTH: fucked-up in London, try Hong Kong." The Asian betting industry offers no downgrade in opportunity, however. A lanky man of twenty-eight years, Brodkin has not enhanced his London complexion out in this port. He sits at his desk in the office of Sportradar, an aspirant of the betting world, his head filled with an inflow of minutiae that underpin the fix. There are analogues of Brodkin, a handful of young men, working all around him on large computer screens of mass data and live-streaming soccer games.

This Sportradar regional office is located in the Wan Chai neighborhood of Hong Kong island. Wan Chai is the place of prostitutes, and nightclubs, and long nights. Wan Chai is a place of lower morals and louder money than the other Hong Kong neighborhoods that draw Westerners, a perfect location for Sportradar, deep in the moneyed unseemliness of Asia's most cosmopolitan enclave. This is a perfect place for the business of detecting manipulation.

A whiteboard hangs along two walls of the Sportradar office. On it is a diagram of the overlapping and interrelated international match-fixing syndicates. The chart lists known or suspected major participants, drawing the connections between them. This schematic is something you would expect to see in the situation room of a police station, not in the new office of a company at the forefront of technology. Match-fixing is such a complex business of ever-changing data and code that sometimes it helps just to tape things on the wall.

A black-and-white picture of Wilson Perumal hangs in the middle of the whiteboard. In the image, he wears a salt-and-pepper goatee, and his head is cast down at an angle, as though he is contemplating figures, or his complicated future. A passport photo of Dan Tan is taped to the board near Perumal. Dan Tan's hair is parted in the middle, and it sweeps down either side of his forehead. He looks like the sort of person who will retain his youthful looks well into middle age.

The writing on the board is color-coded: green for players and refs, blue for fixers and runners, black for financiers and syndicate figures. The front companies of Perumal and Santia are listed in one corner of the board, below a box of tainted referees. Another box includes the words "Triad," "Camorra," and "Russian." There are bubbles with text, connected to each other by multicolored lines of ink. Across the sprawling diagram, there are mentions of activities in Italy, Hungary, Croatia, Belgium, Vietnam, Finland, Guatemala, El Salvador, Togo, Belize, Mali, Spain, Bulgaria, Japan, Zimbabwe, Liberia, South Africa, Bolivia, Colombia, Thailand, Malaysia, Venezuela, Moldova, and Bahrain. Although the board is large, covering two walls, you get the sense from looking at everything that is written on its surface that it is, in fact, small. "Are these linked?" someone has asked in blue ink.

Dany Jay Prakesh and Jason Jo Lourdes are listed in blue, as are "Chinese runners," while Christopher Musonda is in green. There is a photo of Ante Sapina, smiling casually, and one of Rajendran Kurusamy, who is out of prison and, they say, fixing again. There are separate listings of tainted clubs, tainted matches, known fixing scandals. Above the entire diagram are the words "How is Sergei Ussoltsev connected?" The ink on the board is erasable, unlike the mark on the game.

While Eaton's operatives burrowed into the betting worlds of Southeast Asia, his mind kept returning to Hungary. Eaton wanted some answers. Had Perumal fixed World Cup matches? What else did he know? Perumal had talked to everybody else— how could Eaton make Perumal talk to him? Since their late 2012 email exchange, the two had not communicated. In April 2013, Eaton reached out to him again, in an email asking for information about his activities in South Africa. Perumal responded:

> Mr. Eaton i hate to sound rude but the fact is you are
> now employed by a country that had corruptly won

the rights to host the World Cup in 2022. . . . i am sure you will agree with me no one in the right frame of mind will vote for Qatar to be the host World Cup when the temperature is 42 to 45 degrees Celsius in June. Furthermore why do FIFA big shots get a a [*sic*] slap on the wrist. Instead of being judged by a judiciary these crooks face an ethics committee and usually resign even before judgement is passed. . . . I am just a cog in a wheel. . . . These guys are the real crooks the law should go after. People who abuse their positions are the real thieves. I hope you will focus your energy on more useful assignments and do what you can do help eradicate or at least educate players on the consequences of accepting a bribe.

This generated an immediate reply from Eaton.

Dear Wilson,

There are far better ways of getting exercise than jumping to conclusions. . . .

It seems to me that you don't really see either the big picture or the deeply personal one either on match fixing and betting fraud. It is probably because you have been living a fairly difficult and dislocated life for some time now, and maybe you have never had peace of mind.

I also hate to sound rude Wilson, but the big picture is that international sport is in a grave state because of criminals like you. Football particularly is in an urgent need of a truth commission of some kind to cleanse itself, and shed the greedy, unscrupulous predators that circle it. At the personal level it is not you who has spoken to young players in Africa, Central America and South East Asia about what you and your colleagues have done to their dreams and aspirations, and to some of their former heroes.

You may think this too saccharine a view, but it is that youthful and sweet naivety that gives hope in this difficult world, and for some of them you took that away. And for some of those, brutally. Perhaps you have never enjoyed that sweetness and naivety yourself. You spoke to them offering money and rewards to be corrupt. I speak with them to stop people like you. That's our difference.

I felt you would like the opportunity of redeeming yourself genuinely, with real regret and emotion, not merely acting out of revenge and a crushing inevitability driven by being caught and under control. . . . Your damage, your legacy to date, is horrible. If you want to keep it that way or can't see a way out, so be it.

No Wilson. I will not leave you alone. I will not write to you again, certainly.

But be sure that whatever I can do to either pry the full truth from you about what you, and I repeat, you have done to the credibility of football and to young players, or to see that you are properly judged in these other jurisdictions where you have done far more far damage than you have done in Europe.

I expect one day to meet you sitting across a room from me, a court room.

In Asia, Eaton's operatives had encountered their limits. They weren't representing FIFA anymore. They weren't strong-talking fixers. They were now trying to penetrate the multi-billion-dollar sphere of Triad-controlled gambling—a much more difficult task. They progressed only so far. They were not Asian. They couldn't mix in, get lost. Information came slowly to them.

"It's a harder nut to crack," Eaton says. "We are getting to where the money is. And when you get to where the money is, you begin to worry the people who control it. The match-fixers have to be up front. The very nature of it is entrepreneurial. It's a necessary part of the process. It's not a necessary part of the process of betting fraud. This is an underground activity."

In Indonesia, Eaton's operatives instigated a scandal. Two of the most powerful business and political families in the country were rivals in all things, especially in soccer. They owned teams in separate leagues. There was suspicion that each family had been fixing games in order to embarrass the other family's club. One of Eaton's investigators had compiled a list of suspicious names and activity, then handed this to an official from the Football Association of Indonesia. The official leaked the list to local media, which broadcast its contents, attributing the information to Eaton's operative. He was exposed. Within weeks, the operative pulled out of the investigation.

"It's a multilayered beast," Eaton says. "Whereas match-fixing is multiganged, gambling is multileveled. And far more disciplined. And even more international than match-fixing, through its use of technology, and the mixture of legal and gray bookmaking. It is a multilayered, highly disciplined criminal undertaking."

As Eaton's team experienced increasing difficulty in cracking the betting wing of the syndicate, Eaton's mind began to wander once more. He sat on his Venice canal. He looked across the water to the villa that was identical to his own, except that no one lived there. Eaton was alone, and he felt it. He was tired of the travel. He was tired of battling police who didn't share his understanding of the importance of fixing, and the need to cooperate internationally to defeat it. He was tired of the small-time play, the investigations that led to the arrests of players and referees. He doubted he would get a chance at anything bigger. On September 6, 2013, Eaton emitted a tweet: " 'Zero Tolerance'—words relevant only in action. Light suspensions & legal gobbledygook. Perumal partying in Budapest! What's wrong here?" Could it be so bad to have Venice all to yourself? Only if you want the real thing.

Oscar Brodkin says, "It's been a busy day. We had four fixed games come through last night. There was this one from Slovakia. SBO and Singbet eventually pulled the game. IBC refused to offer it." The screen in front of Oscar Brodkin displays FK

Senica versus FK DAC 1904 Dunajská Streda. This is the Fraud Detection System, and the graph of proof that it provides. Suspicious matches are listed in red. But that is only one thing that the system can do. The system compiles lists. There is a list of the countries most prone to fixing. When Brodkin clicks on a listing for Albania, 135 suspect matches appear in red.

"We can even search by player," Brodkin says. On-screen is a list of the most questionable players in the system. Each one is given a "cumulative fraud score." The number-one player here is named Nicola Ferrari, in Italy's Serie B.

The system tracks the movement of these suspicious players. Brodkin mentions Dinaburg, a Latvian club rife with fixing. In October 2009, UEFA forced the club to disband. "Some of the players went down the street to another team," Brodkin says. "They started fixing again." The system is able to select a suspicious club, scan its roster, then monitor the movements of all of those players should they sign on with another outfit. "When a suspected player moves to a new club or a new country," Brodkin says, "we get an alert."

He points to the screen. Several Zimbabwean players have recently begun playing in Cyprus. "Suspicious," he says. Among the many windows open on the screen, there are reports on an English club that plays in Conference South, a small team called Dartford. The screen provides the endless data of the hundreds of bookies that Sportradar tracks, and all of this adds up to a clear designation of who is fixing and where they are doing it. In the lower right corner of the screen, there is collected data on a club called the Southern Stars, which plays in the obscure Victorian State League, in Australia. "There's something going on there," Brodkin says, pointing a finger at the Southern Stars listing on the screen. "I've never heard of this club." No team is too unimportant to escape notice. Not to the international bettor, not to the syndicate, and therefore not to Sportradar. Brodkin's business is the fixer, finding him hiding in this confusion of data. When he appears, he is unmistakable. "That's the great thing about tracking match-fixers," Brodkin says. "They get greedy and do crazy things."

CHAPTER 37

It was early September, another hot evening in Doha, when Chris Eaton's phone rang. Someone was calling from Australia. Eaton didn't want to wake his wife and infant boy, so he stepped out onto the balcony. Overlooking his "fake Venice," he answered. Two cops were phoning from Australia's Victoria Police, his old department. They weren't interested in reminiscing. They explained that they had recently begun an investigation into a series of manipulated soccer matches in Melbourne. All of the games involved a club from Australia's second division.

Eaton listened as the cops described the bits of information that they had so far collected. Wire transfers. Player contact with suspected handlers. Provocative telephone dialogue. And a dreadful win-loss record for the club in question. Eaton had heard it all before. He told the cops that they were on the right track, following the proper instincts. But he understood from their line of questioning that they had no grasp of international match-fixing. How could they have? Eaton well knew how the crime of fixing could confuse the uninitiated, just as it had confused him at the World Cup in South Africa. Over several hours, he meticulously explained the greater global context, how fixing and betting fraud operated symbiotically, and who was ultimately in control. As was normally the case when Eaton pulled back the curtain, he listened as the Victoria cops worried over the morass before them.

The conversation drew to a close. One of the investigators had a final question for Eaton. And now the tables turned. Now it was Eaton's turn to be shocked. He instantly realized that events in Melbourne amounted to no simple case at all, but perhaps the most significant match-fixing investigation there had ever been, the one that might finally spur governments to action.

At the close of the 2012–2013 season, the East London club Hornchurch was relegated to the Isthmian League's Premier Division, the seventh level of English soccer. Hornchurch had posted a record of 11-20-11. Sportradar had recognized suspicious betting patterns for some of Hornchurch's games, and the company trained an eye on the club's activity on behalf of its client, England's Football Association.

On June 22, 2013, a Hornchurch striker, Lewis Smith, signed with Dartford, in Conference Premier, England's fifth division. Sportradar technicians, on alert to all activity related to Hornchurch, watched Smith move. Just two weeks later, curiously, he moved again. On July 8, Smith left Dartford for Australia, joining the Southern Stars.

Located in the southeastern suburbs of Melbourne, the Southern Stars played in the first division of the Victorian State League, which was the second level of Australian soccer. Considering how Australian rules football dominates locally, this second division of Australian soccer is equivalent to England's third or fourth division, or lower. Players in the Victorian State League are amateur. They do not receive salaries. They play on neglected fields, before sparse crowds that often number in the teens.

Yet a group of English players had found the opportunity to play for the Southern Stars attractive enough to relocate to Australia. A handful of English players joined the Southern Stars roster last season. They had little positive effect on the club's fortunes. By the time Lewis arrived, the Stars had played twelve contests, losing ten and drawing two. On the Stars, Lewis joined his former Hornchurch teammates Reiss Noel, a defender, and Joe Woolley, a goalkeeper. In the ensuing five matches, the Southern Stars drew once and lost four times, three of these losses resulting in 4–0 scores. After seventeen matches, the Southern Stars had a record of 0-14-3.

There may not have been many spectators at the Southern Stars matches, but Sportradar was watching. Employees noticed suspicious odds and line patterns in the Asian handicap market for Southern Stars games, the betting-world evidence of match manipulation. Sportradar had recently signed a contract with Football Federation Australia, and shortly after Lewis's arrival, company executives notified their new clients of this suspicious activity. Sportradar had done this before, alerting clients to suspected fraud. In almost all instances, the client federation had chosen to handle the manipulation internally, avoiding public scandal. Instead, officials at Football Federation Australia notified the Victoria Police.

The state of Victoria was one of few legal jurisdictions in the world that was prepared to combat a live match-fixing ring. Because of its global isolation, and the fact that soccer was underdeveloped domestically, Australia, as far as industry experts could tell, had yet to be inflicted with match manipulation. But from numerous international news reports, Australian police and prosecutors understood that match-fixing was a blossoming phenomenon, likely with time to immigrate their way. The local government passed legislation that classified match-fixing as a felony, with a maximum penalty of ten years in prison. While elsewhere in the world, police and prosecutors hardly knew how to charge fixing suspects—with money laundering? business fraud?—Victoria had made things plain. And the severity of the sentencing guidelines automatically relaxed the probable cause that police would be required to show the judiciary in order to enact wiretaps and other invasive surveillance techniques. This would accelerate the process of evidence gathering. The English players didn't know it, but they had flown south into a trap set especially for them.

After speaking with the football federation, then consulting with Sportradar, Victoria Police investigators combed through the immigration records of the Southern Stars players. They were able to establish a group of English imports who traveled together in and out of Australia during the season, on vacations to Bali and assorted Asian destinations. They easily located photos that the players had taken of themselves in nightclubs and on beaches in these places, then posted to Facebook and Instagram. Along with data that Sportradar had provided, this was enough to gain legal approval to record the communications of the Southern Stars players and staff.

When cops began listening to phone conversations between the players, their coach, and several other unidentified men of Southeast Asian descent who were hanging around the team, they realized that they were in over their heads. They knew that they were on to something, and possibly something big. The first cops who engineered a successful case under Victoria's new match-fixing statute would be due professional recognition. But the investigators working the Southern Stars case didn't know how fixing worked and how betting markets could be manipulated. They wanted to be certain that they were focusing on the right information. In need of greater context and understanding, they called an expert.

Chris Eaton told them all he could. But the cops wanted to know more. One of the investigators asked Eaton: "Do you know who 'the king' is?" The cops said that while they listened to phone conversations, they heard the coach, the players, and the handlers frequently refer to someone as "the king."

As Eaton considered this information, one of the cops added: "There have been a lot of calls to and from a number in Hungary."

Eaton's mind cleared. He had only one thought. "The king?" he asked himself. "The *kelong* king? It's Perumal. It has to be. There could be no one else. He's still fixing. We've got him."

The deputy superintendent of the Victoria Police, Graham Ashton, is a large, ungainly sort of man, bald and stern. He made a name for himself during the investigation into the 2002 terrorist bombings in Bali, which killed 202 people. Ashton supervised the reconstitution of an obliterated SIM card from a phone that the bombers had used to trigger one of the deadly explosions. This piece of evidence led investigators to the perpetrators. Ashton brought a similar thoroughness to the Southern Stars operation.

Scrutiny of financial records revealed that the Southern Stars players had received wire transfers of less than $60,000 in total, lower-division players traditionally costing little to control. Telephone intercepts suggested that the Stars coach, Zia Younan, augmented this sum at the conclusion of each successful fix. Investigators learned that Younan had approached Southern Stars management in October 2012. Younan had played professionally in Australia and had since transitioned into coaching. His proposition to the Stars: he didn't want a salary, just full control over the roster. Like many financially strapped clubs and federations across international soccer, Stars management readily agreed to a the proposition, dubious as it was.

Ashton allowed the games—and alleged fixes—to continue, in order to collect evidence that would impress a judge toward conviction. Investigators posed as fans on the sidelines, using directional microphones to pick up chatter between the players. Cops sank microphones into the soil of the playing field. They even placed microphones in the frame of the goalpost. During a match, they listened as one of the defenders told Joe Woolley, the Southern Stars goaltender: "Let the next one in." As the club continued its downward spiral, the players were unaware that police were cataloguing their every move. "There was some interesting body language," says one of the investigators. "You're not supposed to high-five when you lose."

The goal of the fixes varied, depending on the betting market. In some matches, according to recorded conversations, the fixers planned a draw. In one match, they instructed the players to construct a scoreless first half, then allow four goals in the second half. When this game failed to adhere to the plan, ending 3–0, the phone lines activated. The impetus for the calls, according to investigators, was Hungary.

Wilson Perumal would have had a good reason for cracking the whip. Investigators claim that he was betting up to $500,000 on the outcomes. Ashton estimates that the fixing ring cleared roughly $2 million during the Southern Stars operation.

A significant payout came on August 18, when the Southern Stars faced Northcote City, a club from Thornbury, a community on the outskirts of Melbourne. When the two teams took the field at John Cain Memorial Park, in Thornbury, the home club's fortunes were the opposite of the visitor's. With a record of 10-2-5, Northcote City was in first place in the Victorian State League. It would finish the season 13-2-6, winning the league championship. The Southern Stars were a heavy underdog.

At the start of the match, the Asian bookmaker SBOBET listed Northcote as the favorite to win by three or more goals. On the Asian totals market, the over-under was given as 4.5. As the match progressed into the first half, SBOBET altered the line, the odds, and the over-under to reflect the game conditions and the action it was receiving, as was customary.

There was no scoring through the early portion of the first half. Naturally, the odds of a Northcote City win by three or more goals increased, as there was less and less time for the club to reach this total. Likewise, the line changed. But the lines and the odds increased too quickly than they logically should have.

In the twenty-sixth minute of the match, SBOBET was listing the Southern Stars as heavy favorites to avoid defeat by two or more goals. This meant that if the club lost by one goal, or if it drew or won the match, then this bet would be a winning bet. At the same moment, the odds for the over-under, the line of which had dropped from 4.5 to 3.5, were suspiciously distant from where they should have been. Instead of 1.96 (a 51 percent likelihood), which were the logical odds, the under traded at 1.34 (74.6 percent). What did this mean? It meant that someone was placing so much money on an outcome of three goals or fewer that SBOBET had altered the odds so that a winning $1 bet on the under would pay only 34 cents.

This discrepant tendency persisted throughout the first half. The starkest example of it appeared in the forty-fourth minute, just before halftime. The match was still scoreless. By this point, SBOBET had reduced the over-under line to two goals even. The odds on the under now stood at 1.25 (an 80 percent likelihood), instead of the logical 3.09 (32.4 percent). SBOBET was continuing to reduce the odds in order to protect the book against the heavy action it was taking on the under bet. At odds of 1.25, a $1 bet would win just 25 cents.

By halftime, the betting that had initiated SBOBET's odds movements was complete, as the second-half lines and odds adhered to norms. When the match ended, the Southern Stars and Northcote City playing to a 0–0 draw, the betting of the early first half paid out at the high odds under which it was placed.

Not every Southern Stars fix came through. Several Victoria cops claim that they listened in as Perumal scolded the players by phone. Investigators say that Perumal frequently phoned Segaran "Gerry" Gsubramaniam, a Malaysian who allegedly served as the local project manager, and that he routinely reprimanded him for bungled matches. Investigators also claim that Perumal spoke frequently with Younan, dictating the lineup for matches and providing specific instructions and strategy. Investigators traced calls to Jason Jo Lourdes, Perumal's old friend, and a man named Krishna Ganeshan, an established member of the Singapore syndicate. Information from Australia's Department of Immigration and Border Protection revealed that these two men traveled to and from Australia several times during the Stars season. Chris Eaton learned that Ganeshan spent long spells in Brunei, where he met with Crown Prince al-Muhtadee Billah. The prince had once played goalkeeper for Brunei's Duli Pengiran Muda Mahkota club, and now he owned it. When Eaton learned of this development, he directed his subordinates to research ties between the Singapore syndicate and the Brunei royal family.

However, Eaton understood what was of more immediate concern, and the most important thing of all: proving that Perumal had orchestrated, remotely, a match-fixing ring while he was under the protection of Europol and the Hungarian government. If this was true, it would have a major impact on the fight against match-fixing.

CHAPTER 38

MELBOURNE, SEPTEMBER 2013

It is a crisp Melbourne morning as Eaton makes his way down Collins Street, phone in hand. He is typing an email to Ron Noble, describing the morning's events. Arrests began at 5:30 A.M. Graham Ashton's cops raided seven properties across Melbourne. They arrested Younan, Gsubramaniam, and six Southern Stars players. One player had picked up a woman at a bar the night before and gone back to her place. Police experienced difficulty tracking him down. Lewis Smith, the impetus for the investigation, had left the Southern Stars and was safely back in England.

Eaton dispatches the email and places his phone in his jacket pocket. "Noble will puff out his chest when he tells Wainwright," he says, referring to Rob Wainwright, the director of Europol. "Noble will love to say, 'Hey, we have Perumal.'"

That evening, the taxi cruises St. Kilda, through Eaton's old haunt, his first posting with the Victoria Police. He points out the police academy that he entered at seventeen, shortly after his brother's death. "That's where I studied as a cadet," he says. A crowd of local TV crews has gathered around the Victoria police station on St. Kilda Road. Eaton waits also. A small paddy wagon pulls up along the curb. "I once stuffed eighteen people into one of those," he says. "It was a loud party, 1972. All Kiwis. We arrested them when they didn't want to turn down their music. My lapels were torn off. A great fight."

Ashton's cops have spent the last eighteen hours leaning on the nine people they have arrested. It is nearly midnight, and most of the players have already admitted to their involvement in the Southern Stars ring. Gsubramaniam has admitted nothing. Younan also denied involvement. When police showed him compelling evidence to the contrary, Younan burst into tears. He cried for nearly fifteen minutes, his body heaving.

A judge has been summoned to officiate an impromptu bail hearing, and this is held in an unceremonious conference room on an upper floor of the police station. A few reporters are in attendance, but that's all. Police lead Noel and Woolley into the room, one by one. They sit in a chair before the judge, calmly answering his queries. It is difficult to tell what worries the players more, the court or the syndicate. As soon as Noel and Woolley make bail, they barricade themselves in their rooms.

The revelation of Perumal's alleged involvement in the Southern Stars fixing ring underscored the laissez-faire approach that international law enforcement applied to fixers, rather than to players. Fixers and their financiers were more difficult to catch and prosecute than the players who served as their pawns. But the news of Perumal's alleged activity was too galling to ignore. Here was a fixer who was allegedly perpetrating crimes while in the control of law enforcement. Only three days after the arrests in Melbourne, as if in response to the news, Singapore finally spoke.

Singapore police arrested fourteen people in match-fixing raids. Investigators claimed that between 2008 and 2011, this collective had manipulated nearly seven hundred games, mostly in Europe, including World Cup qualifiers and Champions League matches. Among those arrested were former Perumal associates Anthony Raj Santia and Gaye Alassane. Singaporean authorities were cryptic about their haul. But anyone who had been following the international match-fixing story knew what Singapore police meant when they announced that they had captured the "mastermind" behind the entire operation.

The maneuvers that Dan Tan had initiated in Rovaniemi had now finally come back to haunt him. Dan Tan was in jail, and for an unknown amount of time. Singaporean authorities quickly released ten people of the fourteen arrested. But Dan Tan and Anthony Raj Santia were among four held under a section of the Singaporean penal code that allows the state to detain a suspect—without charge—on suspicion of drug trafficking, money laundering, immoral living, or organized crime activity. It was unclear if Singaporean authorities were prepared to charge Dan Tan, release him for extradition to Italy, or simply hold him indefinitely, which was a permissible result under local statutes. Whatever the outcome in Singapore, these were important days in the fight against match-fixing. The tide appeared to be turning.

The night of the arrests, Chris Eaton can be found in an Italian restaurant in Melbourne. Three of his daughters have joined him, two of them with their husbands. Unlike some daughters, whose attitudes toward their fathers complicate and sour with time, Eaton's girls radiate gratification in his presence, feeding off his spirit and vitality. They spar with him, and he good-naturedly returns fire. But when the topic turns solemn, or he has something he'd really like to say, they all stop with the kidding and turn an ear to his authority.

The restaurant is full, and Eaton's party orders their dishes, mostly pasta. A man ambles around the restaurant, an accordion strapped to his torso, playing for tips. Eaton wants to discuss the Southern Stars investigation. "This case is a wonderful example of organized-crime-fighting techniques," he says, leaning in, speaking in confidence, his hands emphasizing his points. "If they had gone the traditional prosecutorial route, they would have waited months to get this information. And these guys would have been back in England, playing for another team. But it's still not good enough. That is the whole problem with this. They are still looking at it locally. Their stage of heroism is in their own country."

The Australian authorities initiated their arrests while Jason Jo Lourdes and Krishna Ganeshan—Perumal's alleged associates in the Stars fixes—were out of the country. Without having these two in custody, using the threat of a prison sentence to cajole them into divulging their operational details, Eaton fears that it may be difficult to lasso Perumal from Budapest, leaving him free to operate.

"That's the most insulting part," Eaton says. "He sticks up his middle finger. I want to break his fucking finger. You can see how careless he is. It got him some wins. But combined with his gambling addiction, it scared Dan Tan into trying to get him locked away. He's always confessed just enough to keep himself off the plane to Singapore. But he's so stupid. He's gonna spend the rest of his life in prison. I got a good feeling in my gut that the cunt's gonna go this time."

The musician roams over to the table, and a general groan comes over Eaton's party. But not Eaton. He sits back in his chair and watches as the man leans into a rendition of "Que Sera, Sera." Eaton knows the lyrics. He joins the musician in singing them. "Whatever will be, will be," Eaton loudly vocalizes. "The future's not ours to see."

CHAPTER 39

Ralf Mutschke, FIFA's new security chief, lacked finesse in his dealings with the press, especially in comparison with his predecessor. He got off to a rocky start in Zurich. One month into his duties, Mutschke uttered the following public comment: "It's not possible to defeat criminal activity altogether, and match-fixing is clearly such an activity. I hope we can minimize the problem and restrict it. But we won't be able to completely eliminate the problem." Mutschke appeared to confirm what many observers had feared about FIFA, that the organization ultimately considered match-fixing an acceptable part of the global game, simply a new component of doing business.

This rejuvenated the debate. Was anyone accountable for eradicating fixing from soccer? If so, who? "FIFA has a role to play," says Michael Hershman, of the Fairfax Group. "It's an issue of leadership and priorities. Match-fixing is enough of an issue for the reputation of the sport that organizations like

FIFA have a responsibility to address it aggressively. Since Chris Eaton left, I really don't see as much of a concentration on the issue at FIFA. Unfortunately, there has been a loss in momentum. At some level, it's been left to the ICSS."

As Eaton rotated through the global sports and security conference circuit, often in the role of featured lecturer, he proposed a new agenda, one designed to provide sports organization with the tools to eradicate fixing. He cited the need for cooperative international agreements between sports, police, and gambling institutions. He advocated the creation of an agency charged with collecting and analyzing relevant intelligence, a multinational, multiagency body that could provide timely advice to governments and sports bodies. He recommended the formation of a task force designed to disrupt organized criminal groups that are involved in the manipulation of matches and betting. A global fund, administered by Interpol, would provide operational financing.

The plan sounded impractical to many people in sports, security, and government. Eaton admitted that it was ambitious. But he wasn't dealing in fantasy. As a model for this new organization, he cited the Financial Action Task Force on Money Laundering (FATF), which the G7 nations had developed in 1989. Administered by the Organisation for Economic Co-operation and Development, in Paris, FATF had been instrumental in establishing international standards and coordinating governments in the fight against the financing of organized crime and terrorism. Eaton suggested establishing his proposed body through UNESCO, which had developed the World Anti-Doping Agency.

Provocative, recognized as an effective, occasionally capti-vating speaker, Eaton spoke at the European Union, in Brussels. His invitation was a signal that governments were beginning to understand the dangers and severity of match-fixing. In Brussels, Eaton outlined the fundamental reasons for the inability of policing bodies, currently disposed, to defeat it. "Sport is global," he said. "Betting on sport is global. Match-fixing conspiracies are global. Betting fraud conspiracies are global. Transnational organized crime is by definition global. But policing is largely nationally contained. Here lies the root of the problem, and therefore to a large extent the solution. Global sport, globally gambled on, is corrupted by globally roaming criminality. Conversely, there is no equivalent global prevention or investigation."

Eaton's proposed apparatus sounded a lot like the early FBI, a cooperative agency designed to face down criminal enterprise on newly expansive terrain. A cogent, convincing salesman, Eaton sounded a lot like a young, idealistic J. Edgar Hoover, the zealous believer in fundamental restructuring.

Like Hoover, Eaton had his targets, Perumal chief among them. In the aftermath of the Southern Stars arrests, Eaton made the rounds with the press. Victoria Police officials made no public comments linking Perumal to events in Australia. This privilege fell to Eaton, and in repeated interviews he pointed the finger at Perumal, whose free Hungarian living insulted Eaton's belief in the way things should be. "One has to assume from what we know that Perumal is involved," he said publicly. "It is absolutely shocking. It shows how arrogant and how fearless these people are." In private discussions about Perumal, Eaton would go further. "This fucker needs to go to jail. End of story."

Eaton's press tour brings him to New York City. Match-fixing has succeeded in making soccer front-page news in the United States, where the sport seldom rates broad attention. Earlier in the day, he completed an interview with CNN. After several hours speaking with Bryant Gumbel for the HBO program *Real Sports with Bryant Gumbel,* Eaton has the rest of the day free.

He takes a cab downtown to the East Village. It is sunny, one of those perfect afternoons that come to New York in either spring or fall. Walking down East Seventh Street, Eaton reflects on what he has accomplished, how he has managed to emphasize the issue of match-fixing, communicating its importance to international decision-makers. "We certainly have a seat at the table now," he says. That may have to be enough for one man. As he moves swiftly down the sidewalk, and through his thoughts, Eaton admits that the extensive international travel and general pressures of the last several years have grown tedious. He understands that retirement nears. "I'm on the last throw of this dice, mate."

After buying a collapsing doll for his son at a Ukrainian knickknack shop, Eaton crosses the threshold of McSorley's. It's time to blow off some steam at the oldest bar in New York. The barman deposits a beer in front of Eaton. McSorley's is busy for a weekday afternoon, a sunny one at that. The place is noisy with loud voices, as people try to make themselves heard over one another. The shrill sound of clinking beer mugs is all that penetrates the blanketing din.

It is so loud that Eaton doesn't hear the alert from his phone. At some point, however, out of habit, he pulls the phone from his pocket, making sure that he hasn't missed any critical messages. And, in fact, he has received an email of particular interest. His face contorts joyfully as he reads it:

> Hi!
>
> Why do you poke your nose in everything looking for publicity.
>
> You are an ex and not a present FIFA security officer.
>
> You now work for a worthless organization and earning blind salary.
>
> If anyone who has the right to implicate and charge me for match fixing in Australia then it has to be the Mebourne [*sic*] Police force not you.
>
> I hope FIFA will take away the World Cup from Qatar and you will be kicked from your job.
>
> For your information you are not a well liked person among the encorcement [*sic*] departments Europe.
>
> You yearn for publicity and sell vital informations [*sic*] that were meant to be secret.
>
> I wish to see you kicked from your job as nuch [*sic*] as you look forward to my downfall.
>
> Wilson Raj Perumal.

Eaton tosses his head back. He laughs so loudly that, for a moment, the chatter in the bar subsides, and all that can be heard in McSorley's is the sound of a man getting an unruly kick out of something. "Mate, isn't that something," Eaton says. "He just can't keep his mouth shut. It just shows he's not an international roaming criminal. He's a lucky thug." Eaton places his beer mug on the bar. "I love these penultimate acts."

He types an email in reply: Wilson, you're scared, and you should be!

Eaton tucks his phone back into his pocket, then lifts his glass in salute. "The fun of the chase," he says.

CHAPTER 40

The chase would continue. In November, a thirty-two-year-old Singaporean man named Chann Sankaran walked into a pub in Manchester, England. It was his third trip to the United Kingdom at the invitation of an Englishman who called himself Joe MacArthur.

MacArthur claimed that he represented an Indonesian mining magnate who was interested in match-fixing. He wanted to fund a fixing operation on English soil. MacArthur had contacted Sankaran over Facebook, and here they were now in England, working out the details. Sankaran told MacArthur that he could fix matches involving three clubs. He claimed that he could arrange for scores of 2–0 or 1–1 in the first half, with an ultimate outcome of either 3–2 or 4–0, depending on how they would structure their betting. He would facilitate wagers on the illegal Asian market. It would also be easy, Sankaran said, to pay a player £5,000 to take a red card at a certain point in the match, in a typical spot fix.

Several elements of the plan didn't hold up under scrutiny. MacArthur said that his backer was prepared to provide €60,000, but this was not enough to fix a match in England. One would need €100,000 minimum to accomplish the task. And although Sankaran was an associate of the Singapore fixing syndicate, he was a fringe player. He had limited experience with fixes. One more detail wasn't right. Joe MacArthur was an alias, the cover for one of Chris Eaton's former FIFA investigators, who was now operating as an independent contractor.

"MacArthur" had sold his services to the *Telegraph,* and video recordings of his meetings with Sankaran subsequently appeared on the British newspaper's website. Britain's National Crime Agency opened a criminal case, arresting six men, including Sankaran. Also arrested was a man named Delroy Facey, a player agent who had played striker for fourteen different English clubs in his professional career, making it as high as Bolton Wanderers, then in the Premier League. This marked the first case of match-fixing in England in decades. The fixing epidemic had finally reached the home of the sport.

A related revelation may not have been so shocking. During one meeting in Manchester, Sankaran suggested that "MacArthur" perform a search on Yahoo. "You search Wilson Raj Perumal," he said. "*Kelong* king . . . He's the king. . . . He's my boss. Everybody in the world know him, man."

CHAPTER 41

BUDAPEST, DECEMBER 2013

It is quiet in the Muvész café, on Budapest's Andrássy Street, save for clattering plates and the frothing of the cappuccino machine. These noises don't unsettle Wilson Perumal. It appears as though nothing does. The Australians have identified him as the mastermind of the Southern Stars ring. British authorities have compiled evidence that appears to link him to Sankaran's activities in the United Kingdom. But Perumal is as relaxed and as talkative as ever.

"This guy Chann," he says. "This guy, I know him from prison time." Perumal explains that he met Sankaran while he was serving time for credit card fraud. Perumal taught him the rudiments of both fixing matches and defrauding banks. They kept in touch over the years since each gained his release. Perumal claims that Sankaran had begun to dabble in fixing, though unsuccessfully. In April of last year, Sankaran was stranded in Cyprus, out of luck after several matches went against him, living proof that fixing's no cinch. "Then he called me. He said, 'Wilson, I am broke. I only have about fifteen euros in my pocket. Can I come and see you?'" The two enjoyed the nightlife in Budapest. Perumal posted several photos to his Facebook page, he and Sankaran pictured in dance clubs.

A waiter in wire-rimmed glasses appears at the table, and Perumal orders a cappuccino. Then he continues explaining.

In the fall, Perumal says, Sankaran contacted him again, following the approach from MacArthur. Sankaran explained that MacArthur had flown him from Singapore to Manchester, covering his expenses. MacArthur made his pitch, and Sankaran was calling for advice. The numbers didn't add up for Perumal. Just €60,000 for a fix?

"I say, 'Chann, are you fucking this guy?'" Perumal says. "'Because I don't believe he's serious about match-fixing.'" MacArthur invited Sankaran for a second trip, and this time demanded to meet the players personally before handing over the money. Sankaran again phoned Perumal. "I said, 'Chann, you're wasting my time. This motherfucker is not going to give you one penny if he don't see the players.'" This necessitated a third trip, and Sankaran phoned Perumal from England in a panic. He couldn't find any real players to present to MacArthur.

Perumal suggested that Sankaran contact an old associate: Delroy Facey. Facey's mother was Grenadian, and he represented the national team in the 2011 Gold Cup. Fixers compromised this tournament, as they did most Gold Cups, and investigators surmise that the syndicate made Facey's acquaintance at that time.

Perumal knows him, but he claims that he inspires little confidence. "I don't trust Delroy Facey," he says. "One moment, he will say he can do this, the next moment he will say he can do that. He was never good to his word. So I've given up on him." But Perumal thought Facey might be useful for something still. "I said if you want fake players who look like real players, you get Delroy Facey. He can help you. So Chann linked with Delroy."

Sankaran phoned Perumal following the final meeting with MacArthur. When Sankaran said that MacArthur had handed over the €60,000, Perumal was surprised. "Chann started to brag with me, that he's a match-fixer. 'I'm doing this game. I have this team. I have three teams.' I said, 'Chann, you fucker, you went there for three days and you have three teams?' It's not possible." In gratitude for his guidance, Perumal says, Sankaran wired £2,000 to him in Budapest through Western Union.

The night of the first fixed match, Perumal says that Sankaran called him, excited about the bets he had placed on the outcome. In the end, the match didn't turn out the way that Sankaran had planned. There were no more phone calls. The next day, Perumal learned of Sankaran's arrest from an associate in Singapore. "I said, 'Fuck, this is a setup.'"

Perumal tells a convincing story, of a sort that he has told before. He supplies just enough detail to explain his presence in the event—as a disinterested friend—though not enough to pin a charge on him. "Wilson's version is the version of a man treading water and scared for his life," says an investigator with operational knowledge of the Sankaran case. "Every conversation he made to Chann is recorded. He was talking to Chann. He was talking to him in partnership. He's now trying to be smart after the fact. At the time, he was in it. There's no way he steps out of this. They will come and get him."

Perumal says that he has relocated to Debrecen, Hungary's second-largest city, 150 miles due east of Budapest. All the same, where he lives and where he goes, these are not closely guarded secrets. Two men enter the café. They sit several tables away, and they don't say much to one another. They just stick around and pretend not to watch.

Perumal says that he has come to Budapest to meet with his attorney. But he claims that recent events do not unsettle him. "Because they cannot trace down to Wilson Raj," he says. "They can't pin it on Wilson Raj. My attorney is not bothered. I am not bothered. If it doesn't trace back to your name, it's a question mark. If Chann says 'yes, he is my boss,' it's hearsay. It's not enough. It's a very complex thing."

Australia appears to present a simplified picture. By now, Perumal has accepted the fact that Victoria Police investigators possess recordings of phone conversations between himself and those involved in the Southern Stars conspiracy. Confronted with this fact, Perumal again is ready to give a little.

"All right, these guys asked me how the predictions can go, how a match can be done," he says. "My opinion. I've given my opinion. People ask me for ideas. I say, 'Throw some good players in an amateur league. It's not necessary that you have to lose the ball game.' Because people know. How many football games can you lose in a season? If I put five good players on Southern Stars—that's what I told these guys. The idea was perfect. But these were not good players. The person who was there running the show felt that these boys had to lose these games. So at some point in time, they were on their own. I said give me a percentage, and I give you my advice." Perumal pauses. He sips his cappuccino. "When the game is going on, this guy might have called me. 'What do you think? What should we do?' I gave my advice."

Just as Perumal appears to admit to his involvement in the Southern Stars scheme, he pulls back. "But I don't know any of those players," he says. "I've never seen their face. They've never seen mine. This guy, Gerry, I never seen him in my entire life." Perumal claims that a Malaysian syndicate arranged the Southern Stars fix, handling the funding and management of the project. Although he may be dissembling, Perumal understands what all of this must look like to the outsider.

"A lot of people are thinking, 'Fuck, Wilson Raj is dead this time. The motherfucker is dumb.' I'm sitting here having a cappuccino with you, and I say I don't give a fuck, because I'm not involved. Ten people can go and say, 'Wilson Raj is my boss.' But can you substantiate how is Wilson Raj your boss? If somebody caught a conversation in Australia, what number? Is it tracing to who? For the Australian police to approach, they must have some form of evidence. Not a tape recorded, 'I'm listening to this voice, and it sounds like Wilson Raj.' I know they don't have anything. Otherwise, they would have been at the doorstep by now. Whatever happened in England, England can try to extradite me. Even Australia, I'm not worried about Australia. I did not send any money to anybody. It's difficult to implicate me. People can say Wilson Raj is involved, but where's the link? Maybe to a certain percentage, I may have been involved. But whether they can extradite me, I don't know."

There may be more yet to hang on Perumal. Just a week after the Sankaran affair came to light, another incident rocked English soccer. Sam Sodje, an ex-Portsmouth player, told a reporter for the *Sun* that he could organize a ring of players to instigate the handing out of yellow and red cards during a match. The price: £50,000. Implicated in the eventual National Crime Agency roundup was D. J. Campbell, a former Premier League forward with Birmingham City and Blackpool, who was on the roster of Blackburn Rovers of the Football League Championship. The last name Sodje appeared among the Facebook contacts of Odira Ezeh, a Nigerian, whom investigators have tied to the Singapore syndicate and who once worked with Perumal.

A lot has changed for Perumal since last summer, when he was unpopular. Once again, he is a favorite of investigators and prosecutors in Europe and Asia. All he claims to think about is minding his business locally. His girlfriend is pregnant. She is expecting twins, due in the spring. "I'm already a daddy in March," he says. "I am comfortable here. I don't have to move. I don't want to go anywhere. But I don't think they will give me citizenship. With a reputation like mine, I don't think they want a citizen." Where will he end up?

Perumal steps onto the street. The door to the Muvész café closes behind him. He doesn't know it, but great movements are happening behind the scenes. Three detectives from Victoria Police are due in Lyon in a couple days. At Interpol headquarters, they will brief counterparts from Colombia, Finland, Germany, Singapore, and the United Kingdom. Australian investigators are pressing for Perumal's extradition on match-fixing charges. (Even Patrick Jay, from the Hong Kong Jockey Club, visited Lyon, in December, where he lectured Interpol cops about the Asian betting industry. Match-fixing is only growing in importance for policing agencies worldwide.) The key to their efforts is whether or not they can positively identify Perumal's voice on the phone conversations that they recorded as part of their case. Finnish authorities have agreed to share audio files of the interviews that they recorded with Perumal while he was in their custody in Rovaniemi.

The maximum penalty for match-fixing in the Australian state of Victoria is ten years in prison. A sentence of five years awaits Perumal in Singapore. And now British authorities have their own motivations for laying their hands on the *kelong* king. Colluding against Perumal, the police agencies meeting in Lyon are blocking his escape routes. It's not hard to imagine Interpol, Europol, and a quorum of European and Asian governments levying such political persuasion on the Hungarian government that Perumal loses the protection he now enjoys. After time spent in English and Australian prisons, he may end up back home in Singapore, ultimately spending the rest of his life in custody. All for fixing a few games.

But could there be an alternate ending for Wilson Perumal? Australian investigators believe that the Southern Stars operation generated more than $2 million in betting profits. Where the money sits, they don't know, for the betting world remains dim to them. Such a sum could facilitate an exit from Budapest, especially for someone so familiar with crossing borders. As national governments and police forces have regularly demonstrated, once a match-fixer absconds from their jurisdiction, he may as well have disappeared.

The days are short now, in the Hungarian winter. A chill has set in. The sun has gone down early, leaving the street in its lonely darkness. On the sidewalks, people hustle to their destinations, wrapped in winter layers. The traffic light at the intersection turns red, and cars collect in a long line. Perumal zips up his coat, and then he extends his hand. "Okay, goodbye," he says. We shake hands, and I hold tight to his. "Wilson," I say, "someone is waiting across town."

"Who is that?"

"Eaton."

Perumal's face goes blank. "Eaton?" he asks, surprised.

"You want to go see him?"

"Chris Eaton." He says the name like he knew it all along, as though he should have expected Eaton to be here, on his tail.

"He's at the Buddha Bar. We could take a cab over there right now."

A grin crosses Perumal's face. "Chris Eaton," he says again. He takes a breath, thinks about it. Then he speaks. "You know the scene in *Heat*?" The traffic light at the intersection turns green, and the cars pass by in a loud rush of sound. "It's a very interesting scene between Al Pacino and Robert De Niro. Face-to-face. And there's one sentence Al Pacino will say: 'You do what you do; I do what I gotta do.' It's the thing. We both have a responsibility. Chris Eaton has a duty. I will try to beat the system. That's what I'm best at." He pauses. Then he says, politely, "Say hi to Chris Eaton." Perumal turns away and walks into the winter.

ACKNOWLEDGMENTS

Appreciation chiefly goes to Chris Eaton, who alerted me to this essential issue, then expended the effort necessary to illuminate it; thanks also to Fred Lord and Javier Mena. Wilson Perumal earns my gratefulness for his courteous recollections. Without Oscar Brodkin and Darren Small, I would have slipped up while navigating *hang cheng*. Zai Mohamed graciously provided early general schooling.

This book began as original reporting for *ESPN The Magazine*. Commitment and imaginative input in Bristol shaped the notes into a worthy product. Thank you: Chad Millman, J. B. Morris, and Donnie Kwak.

The ideal agent, Joe Veltre, was vital in every aspect; thanks, Alice Lawson, for holding together the operation.

Adam Korn at William Morrow in New York. Rory Scarfe at HarperCollins in London. Indebted to you both for your patience and close attention to the manuscript.

As always, Robert Koenig was there with indispensable advice and guidance. Thanks are due to the following for further professional counsel: Megan Abbott, Neal Bascomb,

ACKNOWLEDGMENTS

David Friend, Rufus Fruit, Tony Romando, Christopher Stewart.

Thanks to Angus and James Roven, and to the Truesdales. Deep gratitude to Dad and Craig, the most dedicated readership. And a bow to Dashenka, for inopportune distraction.